MW01258774

WHAT'S LEFT

ALSO BY MALCOLM HARRIS

Kids These Days: The Making of Millennials

Shit Is Fucked Up and Bullshit: History Since the End of History

Palo Alto: A History of California, Capitalism, and the World

WHAT'S LEFT

Three Paths Through the Planetary Crisis

MALCOLM HARRIS

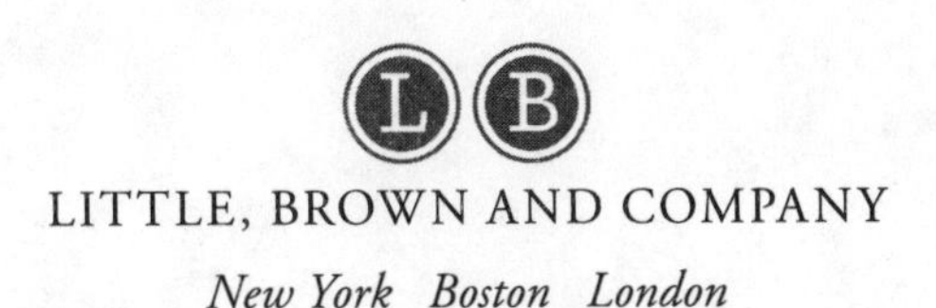

LITTLE, BROWN AND COMPANY
New York Boston London

Little, Brown and Company
Hachette Book Group
1290 Avenue of the Americas, New York, NY 10104
littlebrown.com

First Edition: April 2025

Little, Brown and Company is a division of Hachette Book Group, Inc. The Little, Brown name and logo are trademarks of Hachette Book Group, Inc.

The publisher is not responsible for websites (or their content) that are not owned by the publisher.

The Hachette Speakers Bureau provides a wide range of authors for speaking events. To find out more, go to hachettespeakersbureau.com or email hachettespeakers@hbgusa.com.

Little, Brown and Company books may be purchased in bulk for business, educational, or promotional use. For information, please contact your local bookseller or the Hachette Book Group Special Markets Department at special.markets@hbgusa.com.

Illustrations by Solomon J. Brager

Book interior design by Marie Mundaca

ISBN 9780316577410
LCCN 2024945380

1 2025

MRQ-T

Printed in Canada

For my kid

CONTENTS

WHAT'S LEFT

INTRODUCTION

OIL IS LIFE

"We don't plan to lose money," the Shell analyst told me. When I met him, he was a young man, still completing his transition from geoscientist to bean counter. We talked over drinks as he explained how he ended up working at the multinational oil conglomerate—Shell acquired the small firm where he'd been a researcher—and how he hoped that switching to the energy industry's finance side would provide job security as the oil and gas sector contracted. His role involved analyzing what kind of return Shell could expect from new wells if and when the company is compelled to relinquish them by climate restrictions. Of course, the wells would not then shut down. Rather, the analyst explained, they'd be sold to shadier operators who would continue to pump the wells at lower costs by evading safety and environmental regulations. The firm is incapable of aspiring to abandon its assets; as the analyst said, "We don't plan to lose money."

Since 2017, when I published a book about American millennials, I've had the occasional invitation from corporations to come talk about my work. In 2019, the Shell scenarios team—as in Royal Dutch Shell, which was, at the time, the world's biggest private oil company—offered me £2,000 in exchange for a ten-minute talk and my participation in a group exercise. Its internal corporate think tank was holding a daylong conference about how generational change was going to affect what they call the sky scenario, which the company describes as "a technically possible but challenging pathway for society to achieve the goals of the Paris Agreement." To talk to me about global warming, a giant energy conglomerate wanted to fly me to London from Philadelphia, business class.

I went, though in the end I think I was less help to them than they were to me. I wrote about the experience for *New York* magazine in a feature called "Shell Is Looking Forward," violating the spirit of nondisclosure though not any legal agreement—they failed to make me sign anything.[1] In the months and years afterward, I turned that sentence over and over again in my mind: We don't plan to lose money. We don't plan to *lose* money. We don't *plan* to lose money. Of course they don't. That's the one thing public companies are enjoined from doing, their foundational obligation. To plan to lose money is to willfully violate one's duty to the firm; that's the capitalist equivalent of throwing the game. And yet: Everyone knows that energy giants like Shell cannot exploit all their existing reserves without triggering catastrophic levels of warming. A 2021 estimate published in *Nature* figures that most—58 percent—of oil reserves are "unextractable" if we want even a fifty-fifty shot at keeping the global temperature increase this century to 1.5 degrees Celsius, as was the Paris Agreement plan in 2015.[2] Still, these companies keep investing in finding more oil and gas—products that we can't afford to have them extract and that, once found, they can't afford not to try to extract. Even as the companies have slowed their reinvestment in what they know is a dying industry, in 2022, between its discoveries and acquisitions, Shell added more than one billion BOE (barrel of oil equivalent) in proved *undeveloped* reserves.[3] These reserves are unextractable from a collective human perspective, but humans feel compelled to prepare them for extraction anyway. Though we know that "Water is life," as the anti-pipeline slogan goes, we behave as if oil were more important. We can't stop ourselves; we don't plan to lose money.

It's one thing for a Shell analyst to say something to me over pints at a pub, but it's another thing to confirm the practice. That's what the Environmental Defense Fund (EDF) did in its report on "transferred emissions," which examined data on mergers and acquisitions in the oil and gas industry between 2017 and 2021.[4] What it found was just what you'd expect: large, public, closely scrutinized companies such as Shell selling particularly problematic sites to obscure private operators—sometimes hard to track down at all—easing Shell's way to its climate commitments without taking a planned hit to its accounts. It's not hard to see the collapse of environmental standards at an oil field as they occur. On the contrary, operations quickly resort to burning off excess gas, and the resulting flares and their associated pollution are visible from space. At the *New York Times,* Hiroko Tabuchi reported that in 2021, after Shell and its partners transferred a jointly operated field in Nigeria to a private operator, weekly flaring increased from years of near zero to more than ten million cubic feet of gas.[5] And though the scandal of this sort of activity was global news when compiled by the EDF, the Shell analyst was totally comfortable explaining it to me. It's no secret that companies do their best to make money—the scandal as far as he was concerned would be if they *weren't* selling off these wells, if they were just capping them. That would be betraying the shareholders.

But what about those shareholders? These energy companies are large—some of the largest companies period, for now—and they're not owned by capricious, climate-change-denying individuals. They're owned predominantly by investment funds such as BlackRock and

Vanguard, and those funds are managed on behalf of clients by professionals, and all those professionals believe in climate change, whether they personally believe in it or not, because it's part of market conditions, which makes it part of their job. And though they work for the funds, the funds themselves are owned (unevenly) by a broad swath of Americans, through vehicles such as low-fee mutual and exchange-traded funds. Some activist shareholders have tried to use their stakes in a firm as a platform to persuade larger investors to vote to do better by the environment. The hope is that, with investment already somewhat socialized through these gigantic funds, they can help us effect policy change in the common interest. After all, if you're invested in basically everything—as these funds are—you might be willing to take a hit on your energy stocks if the climate reforms protect your Florida real estate portfolio, for example. At a high enough level, private investment should yield to public pressure, even on its own terms.

In May of 2023, in response to the EDF report, a group called As You Sow organized shareholders in ExxonMobil and Chevron, forcing a vote that would have the companies disclose their transferred emissions as part of their general reports on greenhouse gas reductions. It's a testament to the way large public companies are owned in this economy that the dissidents got almost exactly the same amount of support in both votes: 18.4 percent at Exxon and 18.3 percent at Chevron.[6] That's around the same portion (19 percent) that the activist shareholder group Follow This got at Shell when it urged the company to reduce emissions from the so-called downstream *use* of the company's products, not just their production and distribution—the actual consumption and use of fossil fuels being where the vast majority of (nonwaste) fossil fuel emissions come from, of course.[7] If the great Market doesn't know what companies like Shell are really doing on the ground, it's because it has actively chosen to look away. Rather than automatically compiling and making simultaneous sense of every fact in the world, capitalism knows that sometimes it's better not to know.

If oil is valuable, and if value is the principle by which we organize life on earth, then oil *is* life, even as its continued extraction and combustion also assuredly means death. This bind helps explain why authorities at all levels, everywhere in the world, have struggled to respond appropriately to climate change and its causes. Seeing these authorities as the venal servants of petrocapitalists—though a disturbing number of world leaders are arguably little more than that—is too simple. If it were that easy, we could solve the world's problems with a change in administrative personnel. Unfortunately, people who take command of a society today will be stuck with a genuine conflict between an immediate environmental crisis and a system that generates "a living" for the people they rule. A company like Shell is so intertwined in the US and world economies—from the drivers filling up at its stations to the other companies relying on its industrial products to the pension holders depending on funds that hold Shell stock—that to order Shell to keep its holdings in the ground would immediately destabilize everyday life for many, many people. This fact serves as an insurance policy for the fossil fuel industry, which maintains its high capitalization despite a similarly high level of global social and scientific consensus that fossil fuel emissions must be reduced to zero as soon as possible.

There are a limited number of ways to break this impasse. So far, the world's best minds have not figured out how to fracture the connection between burning fossil fuels and atmospheric carbon or the connection between the increasing concentration of atmospheric carbon and climate change. We are left to snap either the connection between oil and value or the connection between value and life. It's hard to say which sounds harder. Either way, we're going to have to do something about our values.

OIL-VALUE

What do we mean when we say *Oil is valuable*? We don't mean it in the same way we'd say *Life is precious,* the way we'd assert a moral value.* One obvious way we mean it is that you can sell oil for money; it's a product with a price. "Enterprising Child Saves $54 to Buy Barrel of Oil," joked *The Onion* in 2004, above a picture of a little boy dressed as a commodities trader posed next to a superimposed rusty barrel.[8] But the joke only works because, though it's totally believable that a kid could save up $54, what the hell would he do with a barrel of crude? To get the value out of the oil, to actually realize it, someone needs to find a way to *use* it. For even the most enterprising eight-year-old, a barrel of oil is going to be a hassle to move, store, and put to work. We encounter stuff like that all the time—things that, though they have a price, become functionally worthless thanks to factors external to the object itself. *More trouble than it's worth,* as we say.

To make money using oil—as opposed to selling it to someone else, which merely postpones the need for realization—you find a way to plug it into the production process. If the oil could transform itself into profit, there'd be no reason for kids not to save up for their own barrels. But no inanimate object yet devised can use itself, and to realize the oil's use value, you need to mix it with another substance: labor. Like oil, labor power sells for a price, and, like oil, labor power is useful for making stuff, including and especially more money. But also like oil, labor power can't put itself to work alone: It needs to act in conjunction with tools, and materials, and factories, and insurance policies, and warehouses, and enterprise software,

* Though some certainly have said and do say that—see *Anointed with Oil: How Christianity and Crude Made Modern America,* by Darren Dochuk. But I will not be entertaining the dystheistic idea that God consciously nestled fossil fuels under the earth in order to nurture human industry, at least not in this book.

and office supplies, and so on. In our class society, capitalists are those who hold a monopoly over these instruments of production, and they only allow laborers to access the materials we need to make a living if the labor power is *worth the trouble*—that is, if the work's usefulness to the capitalist exceeds its price: the wage.* Instead of capturing the full usefulness of their own work, then, laborers trade some of it to overcome the deficiencies of their class position. For the capitalist, money plus labor equals more money. This un/equal exchange between the classes is the (simplified) basis for the production of value.

This is the most abstract version of where new or extra value comes from, but as far as we're currently aware, there's only one planet in the universe that has ever hosted capitalist production—and done so under a particular set of circumstances. In what Elmar Altvater has called "fossil capitalism," we add fossil fuels and atmospheric carbon into the equation: Fossil fuels plus money plus labor are transformed, via the making and selling of some stuff, into more money plus emissions.† Though in truth none of the elements can propel itself through

* I can type all I want, for example, but to write for a *living* I need to access printers and paper and trucks and warehouses, not to mention editors, publicists, cover designers, sales agents, production managers, all of whom need desks and office supplies, as well as a janitorial staff, which also needs supplies.... But a book is *only worth* acquiring if a publisher can project sales revenue that will exceed its share of those overhead costs, in addition to the author's pay. As you can see, it's workers who pay, via our work, for the very means of production, our alienation from which forces us to enter into un/equal exchange with capitalists. Self-publishing is an option—as are sole proprietorships in general—but that means competing with actual capitalists in a system that's rigged in their favor in innumerable ways. Someone might win, but most won't.

† Andreas Malm writes the "general formula of fossil capital" slightly differently, spinning out the emissions in the intermediate "production" phase rather than presenting atmospheric carbon as a result along with surplus value. See Andreas Malm, *Fossil Capital: The Rise of Steam Power and the Roots of Global Warming* (Verso Books, 2016), 288–92.

the system alone, the system itself moves of its own accord, swelling with every cycle.* Capitalism is production for its own sake.

There are two obvious reasons we wouldn't want this fossil-capital cycle to continue unabated: First, the accumulated atmospheric carbon is rapidly destroying the environment's predictable habitability for all creatures on earth, and that is bad. The second reason is a bit more complicated. Because capitalists are getting some extra value for free out of every minute from every worker they hire—while, conversely, workers have to, in a sense, rent their own opportunity to work from the capitalist—the capitalists end up accumulating value at the workers' expense. Every morning, when workers try to do enough to appease their employers and thus afford to care for themselves in a way that enables them to come back the next day and do it all over again without collapsing from exhaustion, it's their own labor that deepens the class ravine. The rich get richer not thanks to their hard work but thanks to ours. A capitalist can sit at home and do nothing but watch his worth increase *while and because* everyone else comes home at the end of the day worth a little less, having sold their time below its value. That's exploitation, and it adds up on the ledger's two sides. Just as today's excess of atmospheric carbon threatens to destabilize the production process that yielded it, so does the titanic weight of those extorted minutes, hours, years, lives.

Capitalist production is like the deal at the center of Oscar Wilde's novel *The Picture of Dorian Gray,* in which the title character's picture in the attic bears the consequences of his selfish, hedonistic lifestyle, including the ravages of time. Like Gray, we only measure the gold, not the miner's aching back or the mercury in the water system. If you can find a way to bury the costs of your actions somewhere unseen, it's as if

* "Compared with the flows of solar energy, fossil energy is a 'thick' energy source, to the point where fossil energy can easily come to seem responsible for the surplus value produced in a capitalist system." Elmar Altvater, "The Social and Natural Environment of Fossil Capitalism," *Socialist Register* 43 (2007): 39.

they don't exist, and you're free to focus solely on counting your extra value. Capital is then incentivized to enter sectors of production where costs go unmeasured, externalized. Pollution is the classic example of a capitalist externality, and American policymakers have come to prefer improving the market with taxes designed to bring externalized costs back inside the system by assigning them a monetary value to more direct regulation.* These attempts to educate the market have failed, most dramatically in the case of "carbon tax" proposals that sought to bring CO_2 emissions inside market calculus.[9] The market doesn't not know about CO_2 emissions the way we didn't know lead paint was poisoning kids: It's more like the way we don't know where fast-food burgers come from—we do know, but we'd just rather not consider it. As we've seen, shareholders can and do strategically vote *not* to know basic information about their own businesses.

Underlying these ideas about efficient policymaking is the truism that it's counterproductive to try to dictate market outcomes. Rent control and minimum wage requirements, we've been told, will only reduce the housing stock and put people out of work. We can redistribute revenues to the degree to which we can get support for it within the political system, but as far as the market itself goes, all we can do is feed it neutral, true information and trust that we're getting the best possible outcome. Value production for value production's sake, then, shapes the basis for our society, and everything else (including democracy) is downstream. This entire market-fundamentalist perspective is objectively untenable and no longer belongs in the broad conversation about

* "Over the course of the 1970s and 1980s, environmental policy turned away from a moral framework that stigmatized polluters and toward the position that pollution was simply an externality to be priced. Instead of identifying acceptable levels of pollution, the policies began focusing on the most efficient means to achieve previously designated targets. Instead of promoting technologies of pollution reduction, it pushed technologies of market design." Elizabeth Popp Berman, *Thinking Like an Economist: How Efficiency Replaced Equality in U.S. Public Policy* (Princeton University Press, 2022), 9.

relevant responses to the climate crisis, a fact confirmed by its sudden fall from grace to non grata status in both major American political parties as well as on the right and left political fringes. Its only remaining constituency is a small retinue of professional elites who should be treated—to the degree they're acknowledged at all—with limitless disdain for their decades of malpractice. I will not deign to consider them or their ideas.

No matter where in the Oil-Value-Life chain one proposes to intervene, it means collectively grabbing hold of society to make big changes. To subordinate Value to values entails conscious political action, not the flat, amoral maximizing efficiency of economics. This process would be easier if, as it is sometimes portrayed, oil were evil. If, as some seem to think, global warming were a divine punishment for the profligacy with which humans have exploited fossil fuels, then we wouldn't have to make hard choices. We'd just have to share the good news, right ourselves, and the stormy sky would clear to fill with chirping birds. But fossil fuels are not evil; they're an inanimate substrate made of living things transformed via time and pressure. This is what the anthropologist Zoe Todd concluded in the wake of a 2016 oil spill in the North Saskatchewan River after observing its damage to the local wildlife.[10] After all, she writes, "the very pollutants involved in the [oil spill] are themselves the extracted, processed, heated, split and steamed progeny of the fossilised carbon beings buried deep within the earth of my home province."[11] In this way, too, oil *is* life. People all over the world have used petroleum in sustainable ways for thousands of years, to caulk boats in particular, but also for energy. Despite its various unusual properties (and, admittedly, a sort of sinister vibe), there is nothing intrinsically alien or unnatural about the substance. Humanity did not open up Pandora's box or steal fire from the gods; no divine intelligence is collectively punishing us. Climate change is not a fable, and there is no single *self-evident* solution.

Just as we can't stabilize the climate by changing our hearts alone, voluntarily changing our consumption patterns won't halt a system of production for production. The Oil-Value-Life chain persists whether any particular individual chooses to participate, and the system is incredibly good at filling empty slots, as we'll see. "Only by directing our attention away from the realm of exchange do we see the real cause of the climate crisis: a small minority of owners who control and profit from the production of the energy, food, materials, and infrastructure society needs to function," writes Matthew Huber in his book *Climate Change as Class War*.[12] It makes more sense to focus on production—the system's engine and even purpose—than on selling, buying, and consumption. But the spheres are also deeply interrelated. Households consume only small amounts of fossil fuels directly (think cars, stoves, and oil heaters), but, as Simon Pirani notes in his history of fossil fuel consumption, *Burning Up,* "most fossil fuels are consumed indirectly. They are used in the production of materials—from steel and cement to plastics and fertilisers—for industry and agriculture, which in turn produce goods for consumption; as fuel for industry and for transporting goods; for construction; or for military or other state functions."[13] Put that way, fossil fuels are society's lifeblood. Trying to change the mode of production while keeping consumption the same is like trying to swap out a banquet table without disturbing the place settings.

Separating consumption and production, and excluding social processes such as reproduction and waste, are parts of the economistic way of looking at the world, which is in turn part of why we find ourselves in this unfortunate historical situation. If we maintain the separation, we might get the mistaken idea that dramatic change can be isolated to a single sphere, that most of it will be going on behind the scenes as far as we're concerned. That would leave us unprepared for the reality, which is that change is coming for everything, everywhere, because that's where the problem is.

A giant perpetual motion machine that automatically puts people and materials to work and spins off ever-increasing amounts of value is the perfect basis for a society, if you know where not to look. That's why we need a concept that envelops production and consumption and allows us to characterize the larger system: not just production for production but all the stuff capitalist production chooses to ignore, too. I like the phrase "social metabolism," which allowed dissident Hungarian philosopher István Mészáros to analyze the exact kind of planetary crisis we're in now before the twentieth century even ended.[14] Metabolism tends to refer to the way an individual body interacts with the world's matter and energy, but social bodies have it, too. Mészáros describes the concept of social metabolism as "the ultimate framework of reference" for the anchored flexibility of our species. But rather than experience this everything-all-at-once as a single swirling sensation—like a drunk baby crawling through a supermarket—we mediate the world with concepts. These concepts include both the vital "natural requisites" for human life in any society at any time (such as food, water, air, and shelter) as well as the particular patterns of interactions that characterize a specific society (Bed Bath & Beyond). Mészáros refers to these as "primary" and "second order" mediations, respectively. The terms capture the difference between our absolute and relative needs as people. What distinguishes capitalism from the many other overall forms of social metabolism under which people have lived on parts of Earth is that it is not only unbounded but also required to expand. For the first time in history, Mészáros writes, "human beings have to confront a mode of social metabolic control which *can* and *must* constitute itself—in order to reach its fully developed form—as a *global* system, demolishing all obstacles that stand in the way."[15] From a certain vantage point, that's a progressive tendency: Capitalism brings a human community that knows itself as such together under one roof for the first time. But the same uncontrollable growth that makes capitalism unique quickly comes to threaten our relationship with nature, which allows human society to exist in the first place. Capitalism is an *exhausting* form of social metabolism, using up workers and the environment to produce value to produce value to produce value until value replaces air and we all choke to death on it.*

VALUE-LIFE

Can't we just, you know, *stop*? Though changing our consumer patterns is clearly insufficient, we could collectively pick a different kind of social metabolism, one that isn't based on extraction and exploitation. There is no shortage of proposals for alternative systems of values. Suggestions come from on high, thanks to God's representative on earth, the pope; from below, at the World People's Conference on Climate Change and the Rights of Mother Earth, in Cochabamba, Bolivia; and from everywhere in between. Organizers have joined to make big, historic demands about how the world should work differently, whether they're grandparents (Elders Climate Action) or kids (Fridays for Future). How many more people have to believe in what's happening in order to actually make the kind of fundamental changes the situation requires? How many before we can choose to live differently?

Unfortunately, it's not as easy as persuading people to "listen to the best united science currently available," as Fridays for Future demands. That's not just a critique of "scientism" or the tendency to frame social conflicts as objective technocratic questions: Even when people have composed excellent, insightfully human works on climate change, it

* As Nancy Fraser notes, testing the limits of our ecology might be a transhistorical human constant, and many societies have miscalculated and driven themselves and/or parts of their environment to collapse. Capitalism's inherent planetary character, however, means the human community can no longer afford such experiments. See Nancy Fraser, *Cannibal Capitalism: How Our System Is Devouring Democracy, Care, and the Planet and What We Can Do About It* (Verso Books, 2022), 81.

hasn't affected policy. Consider Brenda Longfellow's 2008 short film *Weather Report,* for which she toured Indigenous and peasant communities around the world to get their insights on what was happening. The film is an incredibly successful piece, in both method and content—the kind of necessary scientific work that doesn't always get categorized that way. We have not lacked for this kind of sophisticated climate science, though it too often plays second fiddle to the so-called hard sciences. There's plenty out there, and we can listen and even believe—and so can our policymakers and corporate leaders—but that won't change the facts: Oil is valuable, and value is life. It seems impossible that there should be a reality more real, a set of facts more factual for an earthbound species like ours than the climate crisis, but every day it continues to be the case.

There are two traditional ways to get people to act against their own interests: You can use force or the threat of force, adding a new overriding variable to their calculation of what they want and don't want, or you can trick them, persuading them to replace their interests with your own. Both these methods have a role to play in holding an unequal society together. The police maintain the capitalist monopoly on the means of production using guns and batons; they put razor wire at the border and lock the persistently jobless in jails. Plenty of working-class people are ideological supporters of capitalism, expecting to advance on the system's own terms, and once in a while they even do, proving themselves right, if only at the individual level. But most people don't have to be threatened by the police in order to get up in the morning and go to work and sell themselves short for another day. Nor must we be convinced that our efforts will elevate us to the ruling class. It's mostly something in between, different enough from either one to be worth naming separately.

In his book *Mute Compulsion,* Danish theorist Søren Mau lays out a formulation of that mysterious third force that does so much work to keep the globe spinning. Economic power, he writes, is about control over the configuration of "all the processes and activities needed in order to secure the continuous existence of social life"—which is what we're calling "social metabolism."[16] Because our individual human metabolisms are dependent on this wider social metabolism, we are, in Mau's words, "extremely susceptible to property relations."[17] The fear, when we wake up in the morning and don't want to go to work, is not that the police will bust down the door and pull us out of bed or that we'll miss out on the day's opportunity for advancement but that our conditional access to the means of production will be revoked. We'll be left without a worthwhile way to sell our labor, and as a result our conditional access to primary mediations such as food, shelter, and medicine will also be revoked—not to mention the second-order mediations that feel just as necessary for daily life today, such as internet access and deodorant. Like the God of Revelation with the lukewarm, the capitalist social body promises to spit us out at any moment. That's the fact that's truer than the fact that we need clean air to breathe and water to drink. No one needs to tell us all this explicitly, not on most days; that's why it's a *mute* compulsion.

This economic compulsion is impersonal, and it acts on us and our lives whether we believe in it or not, constantly presenting us with instructions about what decisions are worth making, what kinds of lives are worth living. This is mostly how impersonal historical phenomena act on us persons. Because value production is the basis for our larger metabolism, it comes up in relation to jobs and the consumer markets where we meet our needs (to the degree we can), but it hardly stops there. An uncommitted lover might be "wasting" your time, or a piece of cake isn't "worth" the calories. We use calculating language because we know that somewhere behind our backs there *is* some counting going on about where and how and under what conditions we'll be permitted to make lives for ourselves. That some people manage to close their eyes and buck the system doesn't make the compulsion less real: They're the exceptions who test the rule, and, during the short course of capitalist history, in the clash between anticonformists and economic compulsion, the latter hasn't come up wanting. And when it has, as the artist-musician Laurie Anderson half sings: "There's always force."

It's easy to see how individuals are recruited—volunteer themselves, even—into projects that are hostile to their own existence.* After all, every worker already does the same thing by definition, to a certain degree. Think back to the Shell example. No one *wants* to work at an underregulated oil field that's flaring ten million cubic feet of gas every week. Nor would people choose to live in the world that kind of activity is rapidly creating, not if they got to pick. But if that's workers' only or even best way to fit into the social metabolism, to access the mediations between themselves and the world that make life possible, we know they'll take it. They don't have much of a choice. All the players are then compelled to cooperate: Shell needs to sell the field; the new operator needs to start flaring; and the workers need to operate it the way they're told to. This doesn't require us to think of humans as utility-maximizing automatons; the individuals involved could find themselves overcome with moral qualms and quit—people do it all the time. But that protest presents itself to the system as a personnel problem, not a structural challenge, and such personnel problems mostly solve themselves.

Because this access to the social metabolism is the *ultimate* frame

* Capital recalls a giant creature named the Spirit of Prey in Amos Tutuola's folktale *The Palm-Wine Drinkard,* which mutely compels its victims: "If this 'Spirit of Prey' wanted to catch his prey, he would simply be looking at it and stand in one place, he was not chasing his prey about, and when he focused the prey well, then he would close his large eyes, but before he would open his eyes, his prey would be already dead and drag itself to him at the place that he stood." Amos Tutuola, *The Palm-Wine Drinkard* (Grove Weidenfeld, 1984), 54.

of reference for humanity, it's very hard for another framework, such as environmental sustainability or religious faith or the law, to dig under it. People not only break the law to make a living every day all over the world, they also do so in predictable, calculable ways that describe an impersonal system that incentivizes them to do so. A familiar Drugs-Value-Life chain has steered the totally overmatched illicit narcotics industry to decades and decades of repeated victory against the ostensibly aligned powers of every nation on earth in the so-called War on Drugs. Without other changes, a global War on Fossil Fuels would likely go the same way. To see how, look to the central African nation of Chad.

In the 2013 film *Grigris,* written and directed by Mahamat-Saleh Haroun, we see the kind of incentives that compel people into the illicit petroleum industry. The title character—played by the magnetic Souleymane Démé—is a twenty-five-year-old dancer whose unique moves help earn him enough money to get by. Rather than being impaired by a withered left leg, he incorporates it into his routine, swinging it around his body like a prop. The film opens with him at a club, dancing, surrounded by other young people chanting his name. When the club manager tries to shortchange him for the performance, a couple of his fans make it clear that it isn't worth it to try to mess with Grigris. ("The hat was full, wasn't it?" one asks—the level of exploitation is occasionally subject to these kinds of negotiations.) We also see Grigris working at his stepfather's photo studio, washing bedding with his mother, sewing, flirting. This is a man who makes his way in the world honorably, with individuality and flair—even if he does fake a morning prayer once in a while. But when his beloved stepfather, Ayoub, falls ill and the hospital demands more money than he can raise, Grigris finds himself begging his friend Moussa for the difficult, dangerous job of smuggling jugs of petrol across the border from Chad into Cameroon.

There's no way to turn his leg into an advantage when swimming across a river at night, but Grigris doesn't feel like he has a choice. "Here's the prescription," his mother says, handing him a slip of paper. "I haven't a penny left." When he asks for a job, Moussa tries to turn him away: "You're a good guy. Right? Sell fruit! Shine shoes! Forget it." But because it's worse, smuggling pays better, and so the good guy pleads for a chance to do worse.* What is a customs border or a regulation on lead levels in gasoline to a man whose loved one is about to be spit out by society? Never mind global warming; Grigris is willing to *drown*. "I love that smell," Moussa tells Grigris as he unscrews a jerrican. He doesn't mean the smell of gasoline, of the volatile compounds that make you dizzy; he means the smell of value, of life. For Grigris: Ayoub's life in particular.

Haroun was inspired by the real car chases he saw between petrol smugglers and the police in the Chadian capital of N'Djamena, where the story is set. The film was released in 2013, a decade into Chad's oil boom, which fortuitously began right before America's invasion of Iraq, a period of rapid price increase. The United States became, for a time, the country's largest trading partner, as Chad contended with a drilling consortium led by American companies that preferred to pay minimal taxes and royalties. That is, until America's own shale oil boom reduced the need for Chadian imports, which sank from hundreds of millions of dollars a month to around zero.[18]

Despite having helped power the United States at a crucial moment, Chadians themselves overwhelmingly use charcoal and wood for fuel while still facing the consequences of the fossil fuel industry up close and personal. In one project report, in a section bragging about its "consultation and communication," Exxon featured a quotation from a Chadian "gardener" named Abderrahim Al-Bachar: "I'm not a

* "There's a great future in plastics," *The Graduate* (1967); "Wanna change yo' life? My best advice / A brick of fentanyl," EST Gee, "Lick Back" (2021).

specialist, but I thought the flaring might have affected my crop production. This consultation helped me understand that the flaring was not the problem."[19] After nearly two decades of disputes with Exxon, Chad nationalized its holdings and kicked out Exxon in 2023, transitioning to a closer partnership with, among others, the China National Petroleum Corporation.[20]

One problem with oil is that it doesn't provide a lot of jobs, especially for a country like Chad, which doesn't have a large supply of local petroleum engineers. As a World Bank report on economic diversification dryly notes, "Above 80 percent of Chad's population kept relying on agriculture for their livelihoods and working as subsistence farmers in low-productive informal activities; with Chad's domestic market too small and non-competitive to foster agricultural productivity."[21] ("Go sell fruit," Moussa tells Grigris. "Shine shoes.")

But this is not how progress is supposed to work! As societies get richer, their internal inequality is assumed to attenuate. There's even a graph: The Kuznets curve is a model in which low-paying primary industries yield increasingly efficient manufacturing and a more equal postindustrial economy; it's the premise for whatever confidence capitalists have in poverty reduction over time. However, in a world economy that's already glutted with manufactured goods and the expensive machinery to create them, there aren't many paths up and over the curve. And countries have to compete for access to capital, too. The World Bank report, with a certain degree of exasperation, suggests that Chad increase production of sesame seeds and acacia gum, primary agribusiness exports that might eventually lure foreign investment in machinery, which might eventually produce jobs further along the curve—the global development equivalent of pulling yourself up by your bootstraps.

However, the Kuznets curve is a model, not a law, and no nation—or its people—is entitled to its turn on the trajectory. Examining

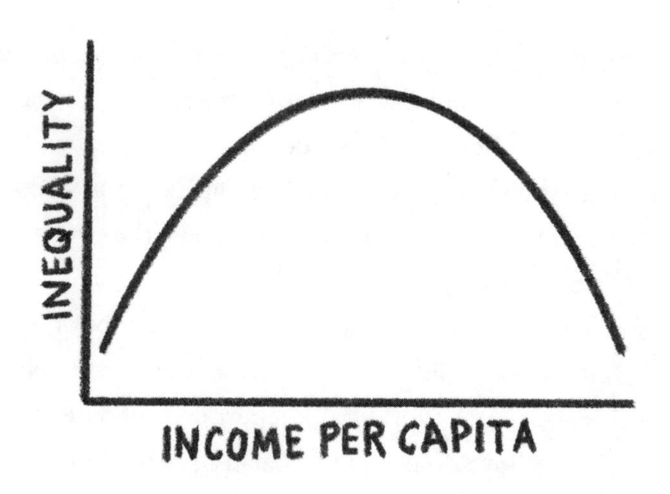

evidence from Ethiopia and Tanzania, researchers concluded that for the nations to grow their manufacturing sectors, they have to compete with better developed Asian economies, and "competing with established producers on world markets is only possible by adopting technologies that make it virtually impossible for significant amounts of employment to be generated."[22] What results is a process that economist Dani Rodrik has called "premature deindustrialization," in which manufacturing's share of employment stagnates and falls. Though, as economic historian Aaron Benanav writes, "Peak industrialization levels in many poorer countries were so low that it may be more accurate to say that they never industrialized in the first place."[23] Haroun depicts this socioeconomic stuckness across all his films, and we see how these historical conditions bear down on Grigris, cornering him into becoming a fossil fuel runner, a regulation dodger, a foot soldier in the war against humanity. He pleads for the opportunity.

Under these circumstances, does anyone expect the Chadian oil industry to contract? No. Rather, the state is building continental pipelines and domestic refineries. A McKinsey analysis of the African oil industry found Chad's reserves "resilient," even in

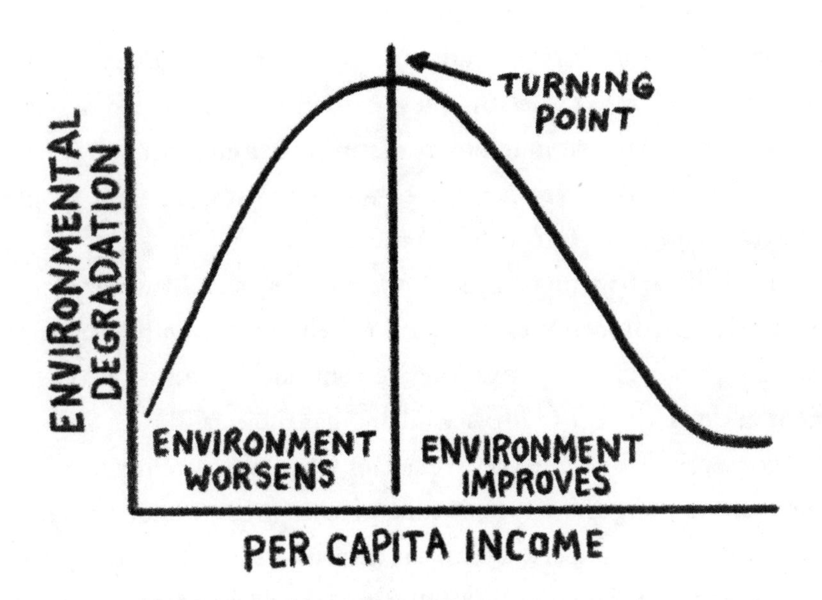

scenarios in which oil demand may have already peaked and barrel prices settle between $25 and $35.[24] Which is to say, as prices drop with decarbonization, Chad's oil will be increasingly worth exploiting as other producers quit. This is bad news, because in addition to an inequality version, the Kuznets curve has an environmental version.

The idea is that production in rich countries tends to be more environmentally friendly than production in poor countries, so the richer countries get, the more environmentally sustainable they become. This version of the curve underlies capitalist environmental philosophy as much as the other one holds down its ideas about international development. But once again, there's no promise that any society—or the larger world—is going in a single direction. Instead, as Andreas Malm argues, investment is incentivized to head toward the curve's *peak*: "Since conditions for accessing cheap and disciplined labour power tend to be bound up with expanding business-as-usual, comparatively high carbon intensity and increased transport, capital will shoot its arrows upwards and backwards, towards the summit of degradation."* Chad's oil industry seems to be becoming more rather than less carbon-intense, which makes its growth potential that much more disturbing from a climate perspective. Workers, as we've seen, are compelled to stop in their progressive tracks and race each other to those arrows.

If Shell isn't willing to *plan to lose money,* should we expect the Société des Hydrocarbures du Tchad to behave differently? The largest economies hope not, because they've continued investing tens of billions of dollars a year to finance oil and gas projects in poor nations.[25] It's nonsensical to blame the workers of Chad for the climate crisis—they rank among the least culpable people in the world, consuming fewer fossil fuels per person than the citizens of almost any other country on earth—yet their vulnerability puts the whole world at risk. That is the consequence of operating our social metabolism according to value production. It's no one's fault, not in particular. We all have to *make a living,* and these impersonal forces dictate which lives are *worth it.* Good luck trying to live one that's not.

REASONABLE ALTERNATIVES TO BOILING OURSELVES ALIVE

Clearly we, the humans of the twenty-first century, have a problem with the way we handle our collective problems. We seem to be acting

* Andreas Malm, *Fossil Capital: The Rise of Steam Power and the Roots of Global Warming,* 339. Further complicating the Kuznets narrative, Lorenzo Feltrin, with others, has introduced the idea of "noxious deindustrialization," in which manufacturing employment declines while polluting primary industries remain, violating the implicit jobs-for-degradation development contract with local populations and generating so-called sacrifice zones. See Lorenzo Feltrin, Alice Mah, and David Brown, "Noxious Deindustrialization: Experiences of Precarity and Pollution in Scotland's Petrochemical Capital," *Environment and Planning C: Politics and Space* 40, no. 4 (January 2022): 950–69; Lorenzo Feltrin and Gabriela Julio Medel, "Noxious Deindustrialisation and Extractivism: Quintero-Puchuncaví in the International Division of Labour and Noxiousness," *New Political Economy* 29, no. 2 (August 2, 2023): 173–91.

out the fable about the frog in the pot on the stove who, only perceiving small increases in temperature, eventually boils to death. But since we're humans, we get the added benefit of being able to have a conversation about the fact that we're slowly boiling to death while we slowly boil to death. And though we share a planet, we don't share a single experience of its destruction. As in Roger Brown's painting of a skyscraper fire, *World's Tallest Disaster,* some people burn while, on a different floor, others enjoy a leisurely cocktail. Instead of the widespread material abundance that many if not most people living in the twentieth century expected to find beyond the 1990s, all we can hope for is to be born on or maneuver our way to the lucky side of ever-taller dikes and border walls. Insofar as that is what we should reasonably expect as the outcome of our present social direction, we are ensuring—at best!—abominable lives for ourselves and our children.

I refuse to believe that we have no alternative to the universal human project's erosion into parochial barbarisms and petty domination. That is an unacceptable outcome, and its giant advancing outline visible through the mist of the near future compels immediate radical action.

This book considers a range of collective strategies for the present moment. It presents what I think is an expansive set of options—from the kind of left-liberal politics embraced under the Joe Biden administration to world communist revolution. But in truth, this is only a slice of the twenty-first century's ideological spectrum. I still believe in the possibility of universal terrestrial progress, which is to say I'm only interested in political strategies for creating a better world for everyone. That rules some things out.

For example, I won't take up the arguments of fascists or nationalists or racialists or other bands who see humanity locked in win-lose contests across lines of permanent difference. Their devotion to division is an obstacle to this book's task, not a solution under consideration.

As I noted above, market fundamentalists are out, their program having been defeated on its own chosen terrain of battle: the marketplace of ideas. I will similarly disregard naive techno-utopianism, which is a variant of market fundamentalism even if it doesn't always recognize itself that way. These ideas are utopian not in the sense that they plan for a better world—that is the goal of this book—but in the sense that they wish away the substance of our actual problems. For example, I will not be considering suggestions that humanity move to Mars, because there is no air on Mars and humans need air to live.

I am also uninterested in the postmodern nihilism that rejects progress narratives and universal politics in favor of cynicism and apolitical withdrawal. Though they make some good points—if the whole progress thing doesn't work out, please don't hold it against me when I show up at your door looking for someone to teach me how to trap small game—our societies and our problems are global, and I haven't given up on action at that level.

The remaining range is a spectrum or even a vaguer cloud, with many individuals and associations locating themselves across a bunch of shifting, hard-to-define axes. Since this book's central consideration is strategies toward collective progress, I partition that cloud based on a few stories about how we're supposed to get from here to there. Though these strategies align more or less with certain labels such as "liberal," "socialist," and "anarchist," I'm going to try to stick to talking about strategic visions rather than political identities. All three traditions have had moments of success in the modern era, moments when their partisans seemed ready to lead everyone into a better world than the one we're living in, and I think all three are worth serious consideration. I'll give short outlines of each strategy, highlighting representative contemporary theorists, policymakers, and organizers.

Realistically, projects that fall into each of the categories are going to continue. My goal with this book is not to convince readers that one is better than the others or that they should convert. I am admittedly a partisan of one strategy, but if I did my job well, it won't be too obvious which one. That's because the situation calls for a sort of metastrategic pragmatism, a strategy of strategies. Without an increased level of coordination, I don't think any of the strategies can succeed on its own. And yet I'll assume from the beginning that success is possible. That's a jump, but it's the minimal one necessary in order to think through our collective problems. We have to be realistic about our realism, too: "On the basis of the evidence before us," wrote the great activist Mike Davis, "taking a 'realist' view of the human prospect, like seeing Medusa's head, would simply turn us to stone."[26] We need more flexibility than that.

Of course there's variation within the categories under consideration, and individuals participate in projects across them: liberals in an antifascist march led by anarchist revolutionaries, socialists in an electoral coalition led by liberals, anarchists in an organizing campaign for increased social services. All three strategies rely on the others, and to the degree each of the three has exhibited plausible progressive leadership, they've found substantial (if not majority) support from one another, as they should expect to. Many anarchists voted for Democrats in 2020, for example, and many Democrats marched to abolish the police—both of which were thoughtful, ethical responses to the respective situations, in my read. But though they all point toward universal progress, there are real differences between these strategic interpretations, and it would do us all well to consider them.

Life on either side of the climate apartheid walls is unworthy of us as a species that possesses a unique combination of abilities, reason, and invention. If, as author Naomi Klein has said, climate change means changing "everything," then what particular thing do we have to change in order to allow humanity to use its collective resources and most powerful skills to address the problems that face us? How do we not only stop "exploring" for new oil and gas reserves but also grab hold of our own productive machinery so that it stops enlisting us to despoil the only planet where we can live? And why is clearing that low bar so damn hard? This book is an attempt to answer these questions, which necessarily means figuring out a strategy for the earth's near- and long-term defense.

CHAPTER 1

MARKETCRAFT

I.

ALICE'S FARM

What could progress look like in this, our actual, disappointing twenty-first century? A YouTube video posted in the fall of 2021 by the user Waffle to the Left offers a surprisingly high-definition picture. In an animation style reminiscent of Studio Ghibli, the video follows a working-age woman as she enjoys a morning on her futuristic farm. A rill turns a tiny wooden waterwheel attached to a mini turbine; a cow luxuriates under a sparkling solar panel; inflatable wind turbines float innocently from flexible tether lines. With the assistance of various robots and a simple cloud-seeding machine, the labor featured is low-key, nothing like the superhuman effort required from today's farmworkers during harvest. One scene shows a diverse, crowded table of happy people about to dig in to a feast of minimally processed food, their automaton friends buzzing nearby ready to serve. Kids arrive on a flying school bus, and in the distance is a vaguely green-looking city. It's a peaceful place, a contemporary glimpse of the kind of lifestyle people have long imagined would someday be our common entitlement. How did Waffle, a Canadian producer whose previous videos—none of which contains original images—struggled to get one thousand views, create such a powerful visualization? The answer is that they didn't. Capitalists did. Waffle just scratched off the brand name.

The video is called "Dear Alice," and it's an advertisement for the yogurt brand Chobani. If you look closely at Waffle's "decommodified edition," you can make out the traces: a drone adds a plastic-looking carton to a table full of produce, and the recognizable single-serving cups are incongruent with the rest of farmer Alice's fridge contents. But that's easy enough to ignore, especially with everything else going on, and that's a low price for the viewer to pay for the rest of the animation. Waffle's explanation for the decommodified version is that we can enjoy it once we expunge the "ugliness of capitalism," but this ends up being the best argument I've seen for the first strategy under consideration: If capitalism can harness technological innovation to generate sustainable abundance and minimize socially necessary labor, it won't matter whose brand is on the yogurt cups. Eventually, as that kind of vulgar competition attenuates, we'll stop printing logos entirely. With the edit, Waffle in effect enacted the green transition. Even if the pierced, tattooed future young residents of Alice's Farm scoff at capitalism, too, the retired capitalists will look at the robots and solar panels and wind turbines—at the world they made possible—with pride, as a culminating success. This is the strategy of marketcraft.

Sometimes those of us on the radical left are loath to admit that many liberals have the same goals as we do: a world in which no one has to work very hard, everyone has their needs abundantly met, and humanity spends most of its time on higher pursuits, whether that means art, scientific research, playing dominoes with friends, what have you. That's the strongest version of the liberal progress narrative as I understand it, and the strongest version is the one that we should spend our time considering.

The neoliberalism critique suggested that liberal politicians joined conservatives in a laissez-faire unpopular front at the end of the twentieth century, but the liberal establishment has shown itself newly enthusiastic about wielding state power over the economy, especially when it comes to the green transition. And if the progressive-minded liberals are right, we can exercise democratic control over our social priorities via market regulation and incentives, shoving firms into competing to build us such an abundance of efficiency and value—including immaterial and hard-to-measure values such as carbon neutrality—that market competition largely ceases to be worthwhile. The market works itself out of a job as the production system nears full automation and as global society, correspondingly, adopts a variegated but universal lifestyle of leisure.

Between here and there is the climate crisis, but as the economist Alessio Terzi writes in his book, *Growth for Good: Reshaping Capitalism to Save Humanity from Climate Catastrophe,* "The only credible way of avoiding a climate catastrophe is to accelerate the development and widespread adoption of 'green' innovations."[1] Accelerating the development and widespread adoption of innovations is what capitalism is supposed to be good at, and we can use the government to program the economy to bring us the right ones. That's what led *New York Times* columnist Thomas L. Friedman to write, in an influential and perhaps even prophetic 2007 essay called "The Power of Green," that "the market alone won't work. Government's job is to set high standards, let the market reach them and then raise the standards more."[2] He called for a Green New Deal to spur one thousand new American clean-power projects.[3]

Yes, as much as everyone would like to claim the Green New Deal label for their program, it's the liberals who called it first, and it's the liberals who have finally enacted something very close to Friedman's program. Trumpeting a wave of private investment in green energy following the Inflation Reduction Act of 2022 (IRA), former director of the National Economic Council Brian Deese wrote that it "has the potential to drive a more rapid and efficient decarbonization of the economy while increasing the supply of clean energy and maintaining the country's competitive edge of stable, low-cost energy."[4] Deese linked to an April 2023 press release from the industry group American Clean Power Association, which announced a $150 billion investment in twenty-seven solar, ten battery-storage, eight wind, and two offshore wind manufacturing facilities.[5] That's not Friedman's one thousand, but it's a step in that direction.

MARKETCRAFT

In contrast to market fundamentalists, proponents of this framework see markets as functions of public policies. Though the idea of a social realm where people exchange goods and services is way older than capitalism, it is not actually the basis for the existence of our species. Rather, it's our coming together as a rule-making polity that allows markets to function in the collective interest and produce a universal rise in the tide of human well-being. If the market becomes detached from our values and metastasizes in places where it doesn't belong—which is to say, when the market ceases to be useful and instead starts making use of *us*—it's up to democratic society to reassert itself and show the price system who's boss.

Bossing around the market in the collective interest doesn't just mean raising wages, decreasing hours, and making work easier—though it does mean those things. It also means exercising the public's prerogative to define the game's rules. Production for production's sake gets results, but there needs to be an asterisk added for the many rules and conditions that may apply. We forbid, for example, literally grinding up human children for Soylent Green Jr.–brand energy bars, no matter how profitable such an enterprise may appear under the market's blinkered, maximizing gaze. At the same time, it's important to avoid simplistic ideological binaries such as "society versus market" and "freedom versus regulation," which presume some original opposition between the processes we're reviewing. That's like saying "body versus kidney"; there can be no production or consumption on a dead planet. Rather, we need a concept that begins from the idea that markets are projects of societies—certainly not the other way around—a project that *subordinates* the former to the latter. Starting that way makes it easier to accomplish the subordination of the market to society in practice, something we know is necessary if we are going to break the Value-Oil-Life chain that's wound through the gate doors, blocking the good paths forward for our species.

To approach the market's management as a metabolic matter, political scientist Steven Vogel offers the useful term *marketcraft,* from which this strategy takes its name. The term is meant to suggest a governmental function equal in importance to the statecraft of international diplomacy. (Personally, I like the resonance with *witchcraft,* from which sociologists Karen E. Fields and Barbara J. Fields also draw their concept of racecraft.[6]) Looking at all the ways the state—and the wider civil society, in some aspects—structures the interchange of economic actors helps dispel the simplistic policy division of regulation/deregulation and its analogous conceptual binaries. At the levels of law, standard, and norm, our democratic institutions build the streets where businesses drive, whether with rules against insider trading, generally accepted accounting practices, or rhetoric about fair play. Or literally, as Vogel points out, when it comes to the trucking market, which operates on top of state-built, state-maintained, and state-patrolled motorways. "The argument begins with the basic recognition that real-world markets are *institutions*: humanly devised constraints that shape human interaction," Vogel writes.[7] If that's the case, then we can use the marketcraft bolt cutters to snap the link between Oil and Value, creating new rules and conditions that devalue fossil fuels from every direction until they are not worth the trouble of pumping from the ground.

When it comes to tackling the climate crisis, most marketcraft advocates have found the hands-off carbon-pricing approach lacking. (It's a testament to Vogel's concept that, as it turns out, fabricating a private carbon market takes a ton of prohibitively complicated government work.) Instead, they've turned to supply-side industrial policy, which means paying companies to actually build the actual stuff that society actually needs. There are a million different marketcraft spells to accomplish this kind of thing—tax credits, purchase agreements, green banks, direct subsidies, and more. In Vogel's thinking, even a rhetorical commitment to long-term investment in clean energy from a financial giant like the US government is enough to shift world markets.

The reality is that, in this capitalist world, most of the direct decision-making authority with regard to the deployment of our collective resources and productive capacity is in the hands of capitalists. The marketcraft thesis is that the state has the power to shape the systematic compulsions they face without us all having to undergo a protracted guerrilla struggle. Capitalists may be the players for now, but democracy can write the rules. Writing those rules to reorient the world economy away from fossil fuels is a matter of immediate concern: In some areas, we can't plausibly mediate private power—by taxing it away, say—before we need to put it to work building the means of green energy production and distribution. Explaining the situation to the capitalist leadership personnel, however, has not motivated them sufficiently, and neither has asking nicely.

We don't have time to wait for the profit hounds to sniff their way to decarbonization, especially when we, the leash-holding public, already know what we're looking for and roughly where to find it. Let's toss a cartoonishly large, juicy steak in that direction and let the hounds off the leash to figure out the specifics—within the strict parameters of their training, of course. The steak is not a reward for good behavior; it's a tool to effect what we need to accomplish.

For many years, Americans heard from economic experts on both sides of the aisle that this "throw a steak" tactic was unworkable. Either we couldn't afford the subsidies or we'd ruin the markets by interfering with them. Market fundamentalists pointed to the failure of the highly touted high-tech manufacturer Solyndra. The firm, which made fancy proprietary cylindrical solar panels, absorbed hundreds of millions of steaklike dollars from the Obama administration only to succumb to competition from China, which scaled up simple panel production, ruining whatever price advantage Solyndra might have been able to project.[8] (It didn't help that the Obama administration simultaneously helped craft a domestic fracking boom that pushed down fossil fuel energy prices.) It was an embarrassing failure for the Democrats, one that nudged the party's mainstream away from Friedman's Green New Deal proposal, ceding the space to democratic socialists within the party such as Alexandria Ocasio-Cortez, who spearheaded a 2019 GND resolution that called for, among other things, a job guarantee and a net-zero carbon economy by 2030.[9] After decades of market fundamentalist consensus, a bit of good old liberal marketcraft can look and feel like socialism.

The COVID-19 pandemic made market fundamentalism practically impossible, and the United States marketcrafted its way to vaccines and economic recovery much faster than analysts expected. Vogel might caution us against talking about government involvement in terms of more or less, since so-called deregulated markets take a lot of regulatory work, but economists and policy advocates who call for openly crafted market outcomes have found an increasingly warm reception after so long out in the cold. Bernie Sanders's economic adviser Stephanie Kelton's 2020 book, *The Deficit Myth,* was a surprise hit, helping set the stage for the Biden administration to present big spending packages.[10] "Yes in my backyard" YIMBYs have crafted a movement around their call to better craft housing markets to increase construction and meet thwarted demand.[11] Even the verboten tool of price controls got new life, thanks in large part to Isabella Weber's research into the comparative marketcraft policies of Russia and China.[12] The pandemic CARES Act crafted the housing and labor markets with an eviction moratorium and so-called super-unemployment benefits.[13] Still, many progressives did not expect much when it came to Biden and climate spending, mostly because the deck seemed stacked against the administration with the human embodiment of coal, Joe Manchin, as the swing vote in the Senate. But by using revenue predictions from a corporate minimum tax, an increase in IRS enforcement, and drug pricing reform, Democrats sold hundreds of billions of dollars in climate spending to the public as an "inflation reduction act"—not as big a steak as the party's left wing wanted but certainly a bigger one than anyone expected to get out of Manchin.

Marketcraft rejects the framework of government intervention in the market the same way a parent might tell a defiant teenager that they can't lock the door while they work with their study partner: *It's your room, but it's my house.* And yet marketcraft, as its name suggests, relies on a system of private capital to fulfill public wants and needs. There is an obvious tension there, a tension that has made it very difficult to fulfill those public wants and needs. Private capital has not proved itself a reliable partner for human society when it comes to the climate crisis—far from it. The capitalist dog has been gnawing on society's feeding hand for as long as most of us can remember. Why squander our newfound confidence with regard to the public's ability to fund the production of our own needs by handing it over to ruling-class exploiters? For the same reason you wash an oil-covered bird with petroleum products.

The moody teenager and hungry dog are metaphors, but the sick bird is literal. Procter & Gamble has long advertised that rescuers use its Dawn soap to clean animals after oil spills. It's a perfect way to present what in 2010 a spokesperson described to NPR's *Morning Edition* as the product's "toughness and mildness."[14] That toughness and mildness, however, is thanks in part to petroleum derivatives—a Procter & Gamble spokesperson gave the percentage at a specifically nonspecific "less than one-seventh," adding, "To say Dawn's horrible because of this, that doesn't make a whole lot of sense, and that's what we're trying to avoid." Dawn has been used to clean oil-stricken birds in particular since it won in a 1978 test of dish soaps held by the International Bird Rescue Research Center (and funded by Chevron), much to the frustration of chemist Martin Wolf, whose company, Seventh Generation, sent an unacknowledged truckload of petroleum-free detergent to help with the cleanup after the 2010 *Deepwater Horizon* oil spill in the Gulf of Mexico.[15] "I think it's extremely ironic," he complained to NPR. "Here we are trying to squeeze every last drop of oil we can out of the earth, and it's despoiling the earth. And we're using that same product that's messing up the earth to clean it up." But if you're a veterinarian dealing with a warehouse of oil-slick pelicans, all of whom will be poisoned to death if they try to clean themselves, aren't you going to grab the most effective detergent you can reach? There are worse things than irony.

GREEN CARROTS

Marketcraft as a practice has no particular political allegiance. Banning lead paint is a kind of marketcraft, but so is deregulating investment banks. Marketcraft as a strategy for universal progress in the face of climate disaster does have a politics, however, and that's what we care about. Once we denaturalize markets and see them as structures people have to build one way or another, we have to ask ourselves what exactly we want our production system to do. Capital says production for production; democracy must craft a narrower path.

How do you produce a negative? More than any particular gadget, the world's people need there to be less carbon in the atmosphere. To have any chance at accomplishing that, as I've argued, we need to break a link in the Oil-Value-Life chain. By programming the market to produce a surfeit of green energy, marketcrafters hope to reduce the demand for fossil fuels as much as possible, limiting the market to specialty products for which we've yet to develop electric substitutes (such as jet fuel) and choking off the industry's ability to invest capital. This plan has already started to work, and the energy giants have been unable to reinvest their windfall profits into oil and gas exploration at the same level they might have in the past.[16] Even they know the fossil fuel future is dimming.

The American marketcraft strategy for climate centers on decarbonizing the electric and transportation sectors. This requires, first of all, a bunch of money. One set of ways the government can craft a market is by offering direct and indirect funding and financing for stuff the public wants to happen. Subsidies are only one way to do it, but they're popular with capitalists for obvious reasons, especially compared to alternative tactics such as fines and bans. That makes subsidies a good thing to foreground when it comes to crafting a class compromise. Citing the CHIPS and Science Act's bid to revive US semiconductor fabrication, the green energy push in the Inflation Reduction Act, and spending in the Infrastructure Investment and Jobs Act—all under President Biden—the Roosevelt Institute proclaimed a "surge of interest in a proactive approach to shaping industries and markets" and issued a "marketcrafting" report in May of 2023, suggesting a few more directions.[17] Among the missions is achieving net-zero greenhouse gas emissions by 2050. "To achieve this mission, the price of non-renewable energy must rise relative to other energy sources," the authors write. "But we should craft public policy to ensure it does so in a controlled fashion, providing an orderly transition and minimizing disruptions for most Americans."[18] That is a good bird's-eye-view description of the marketcraft strategy.

Marketcrafters tend to reject holistic ecological frameworks—metabolic models that verge on spiritual. Climate change is not divine punishment from Gaia; climate change is science. We need to decarbonize the electricity-generation system and swap out our gas-powered tools for electrified ones that can run on batteries charged by solar, wind, hydro, and other clean sources. And we need to build a reliable storage and transmission system that's sufficient to ensure the controlled rollout described in the Roosevelt report. Making sure the rollout is controlled is important not just to secure capitalist buy-in but also to try to nip any sort of antienvironmental consumer backlash in the bud. The state could craft the transportation market by banning internal-combustion-engine vehicles from public roads, but many Americans would likely resist such a policy, a lot, with guns. It's much easier to juice the production of electric vehicles with targeted tax cuts. Plus, this green industrial policy, as it's been called, depending on the conditions attached to the subsidies, has the potential to create a bunch of good manufacturing jobs, further building support for the policies. Cutting green ribbons is solid politics, and we can't be afraid to spend money. By anyone's measure, we're going to need to: Though global investment in energy-transition technologies topped $1 trillion in 2022, the International Renewable Energy Agency says we need to get to $5 trillion a year, fast.[19] Subsidies are a way to prime the pump, so to speak.

There is certainly a Solyndra-type risk in subsidizing green energy technologies, but compared to the risk of climate crisis, it's not clear that making the occasional poor investment even *is* a risk. What's at stake, the Treasury's credit rating? As the economists Mark Paul and Nina Eichacker argue in the *MIT Technology Review,* we need *more* Solyndras: "While Solyndra's downfall received a lot of spilled ink in the media, Solyndra was actually one of only two failures. The other 22 companies repaid their loans, resulting in a *profitable* program overall that helped accelerate multiple green industries in the US."[20] Taking risks on decarbonization tech is good, and if any entity in the world can afford to do it, it's the United States government. But just because the subsidies are there doesn't mean anyone is necessarily going to take them. There's a lot more crafting to go. If capitalists are going to invest many billions of dollars on the basis of these subsidies, they need to be pretty damn sure they're eligible, which means the state needs to offer precise guidance on what it means to produce "zero greenhouse gas emissions," for example, or to pay the "prevailing wage," or to source components "domestically." Still, if haste makes waste, we're going to have to resign ourselves to the idea that decarbonization is a big, costly task rather than an occasion for penny-pinching.

The Roosevelt Institute suggests a way to craft the best chance for these public-private experiments: maintain a steady stock of raw materials that are crucial to decarbonization technologies, including lithium, cobalt, nickel, and copper.[21] In this financialized market, a blip in the price of one of these minerals could take out a whole company—and with it not just our public money but also the deeper investment of time and energy in whatever potential solution the project represents.* Subsidizing the production of materials all the way down the value chain insures against the caprices of global supply networks. In August

* What do we even mean when we talk about these elemental metals in abstract terms? Analysts at Employ America have called for the state standardization of contracts for critical minerals in order to craft effective spot and futures markets like the ones constructed for gas and oil. See Arnab Datta, Alex Williams, and Alex Turnbull, "Contingent Supply: New Benchmarks Can Define and Deepen the Lithium Market," Employ America, September 13, 2023.

of 2023, the Biden administration announced $30 million in support of domestic efforts to refine climate-critical minerals from coal and coal waste, on top of a pair of $8 million grants to the University of North Dakota and West Virginia University to study similar techniques.[22]

The supermajority of government climate funds so far are for stimulating the supply side, to subsidize and finance businesses that make green stuff. But there's something in there for the demand side as well: tax rebates for consumers who buy the right electric vehicles and heat pumps. As with the corporate incentives, we should think of this less as a reward for being eco-conscious and more as marketcraft—a national helping hand to make electric transportation and HVAC systems a bit more affordable relative to the fossil fuel alternatives. It's good politics to have some of the energy-transition subsidies flow into the system through consumers, but there's also a danger in operating on such an abstract level in relation to individual households. What use is an efficient new heat pump in a house with a broken window and a leaky roof? Other policymakers could learn something from the Whole-Home Repairs bill spearheaded by Pennsylvania state senator Nikil Saval, which offers individual grants up to $50,000 for basic home repairs, including but not limited to energy-efficient weatherization.[23] Rather than send a reimbursement check or tax rebate after the fact, a local program administrator reviews applications and hires contractors directly to do the work. That way, applicants don't have to front money that they may not have in order to access the repairs they're entitled to under the law. It's not enough to give away money; even those markets must be crafted with empathy.

Though commentators and the Biden administration itself have embraced the idea that they did more marketcraft than anyone since FDR, the policy has taken a notably agnostic view of what exactly green capitalists are supposed to build. The tens of millions of dollars directed toward mine-waste refining is a drop in the bucket compared to the IRA's $27 billion Greenhouse Gas Reduction Fund (GGRF), which provides grants for so-called green banks that will whip that capital into billions more in lending capacity for green projects. The bill's clean-energy investment and production tax credits are technology-neutral rather than apportioned to wind, solar, and so on. Those are marketcrafting decisions, too, and they're a signal that the state is looking for good clean-energy ideas to support rather than offering them in advance. It's also a choice about how these efforts are to be regulated: If the Environmental Protection Agency were undertaking its own direct green projects with the GGRF, the ventures would be subject to the scrutiny of the National Environmental Policy Act (NEPA) and its dreaded environmental impact report requirement. But private efforts financed by banks that are backed by the government face a lower set of standards. Perhaps, given the daily accumulating consequences of our failure to decarbonize, the emphasis on speed is appropriate.

All this money flowing to corporations, some of which had something to do with getting us into this mess in the first place, does feel a lot like a sellout. Marketcraft leaves the Value-Life link intact, and capitalists will only pursue green opportunities if they are sufficiently profitable by their own standards. They don't plan to lose money any more than Shell does, and that means some rich people will get richer. After decades of austerity politics for the public and pork-barrel politics for bomb makers, it's hard to think about public money as anything other than a prize for rigged tugs-of-war between good guys and bad guys. But the theoretical minds behind the most progressive marketcraft agenda imagine something more radical.

One problem with the strategy as I've described it so far is that, rather than offload our worries onto various companies, the American public could potentially become the world's most overworked debt collector, bound to continually check up on every green borrower lest it run off with the cash. We need a way to fill these private actors with public spirit, tweaking their motives like a little brain parasite. Luckily, investors do that kind of thing all the time. Rather than private equity funds that tend to sweep in and reorient enterprises toward short-term gains, the marketcraft strategy could make use of a *public* equity fund that does the opposite, reorienting enterprises toward long-term ecological sustainability.

In August of 2020, policy thinker and top marketcraft intellectual Saule Omarova published a short white paper in conjunction with Data for Progress called "The Climate Case for a National Investment Authority."[24] In it, she lays out a number of empirical and theoretical problems associated with getting private investors to fund needed green infrastructure. "Individual investors fundamentally lack the capacity to control the broader macro-environment, and their risk-return calculations are driven by their expectations of private profit," she writes. "Their short-termism is, therefore, fundamentally individually rational. Yet, the cumulative result is collectively irrational and tragically ironic: many potentially beneficial infrastructure projects simply do not get funded in private markets, while abundant private capital is desperately searching for profitable deployment."[25]

Irony in the capital markets, it turns out, is a bigger deal than just cleaning birds with petroleum derivatives. A National Investment Authority (NIA) would deploy billions of dollars and various crafting powers to force the markets to see public reason, whether they'd otherwise be inclined to or not.* To accomplish that, Omarova suggests a two-part structure, including an NIA bank (National Investment

* Other commentators have called for a similar structure based on the twentieth-century Reconstruction Finance Corporation. See Kevin Baker, "The Role of Public Capital: A Reconstruction Finance Corporation for the 21st Century," *The American Prospect,* December 5, 2019; John Cassidy, "It's Time to Establish a New Reconstruction Finance Corporation," *The New Yorker,* March 24, 2020; Nic Johnson, "Reconstruction Finance," *Phenomenal World,* April 28, 2021.

Bank, or NIB) and an asset manager (National Capital Management Corporation, or NCMC—or the charming Nicky Mac).

Beyond being a green bank, the NIA could do more than lend to individual projects. It could package worthy public and private infrastructure investments together and offer them as medium- and long-term bonds to the private market, attracting institutional money from sources such as "pension funds, investment companies, insurance companies, foreign central banks." Nicky Mac would take on more adventurous projects, sending cutthroat financial operatives out there to build "nationwide clean energy networks, high-speed railroads and broadband, regional air and water cleaning and preservation programs, environmentally smart and affordable housing programs, systems of job-retraining, networks of public-private R&D hubs, and so on," by any financial means necessary.* The NCMC could take equity stakes in infrastructure projects and even spin them up into new fully public entities. And to pull in private capital, the big brains at Nicky Mac could synthesize all sorts of financial instruments using the federal government's backing.

This synthetic-instrument part is a particularly far-out idea, because it would allow the NIA to directly replace private values with public ones. Omarova gives the example of a localized fund that would return a percentage to investors based not on the profitability of the projects under management but on their general macroeconomic impact in the area. Imagine you open an ice cream shop with Nicky Mac. In this situation, the NCMC pays you a guaranteed return on your investment plus a performance bonus, but the bonus could be per scoop rather than per dollar of profit. The shop may barely break even in an accounting sense, but if your pockets are full and the neighborhood is

* As Omarova points out, the NIA would not be bound by federal-employee compensation limits, allowing the institution to compete with private financial firms for experienced and talented personnel. See Omarova, "The Climate Case for a National Investment Authority," Data for Progress, August 2020, 7.

smiling, what else is there to care about? The government isn't going to go broke—in fact, some Nicky Mac finance witches have cooked up a model that says every underpriced ice cream scoop is leading to an average of three dollars in extra tax revenue somehow, so it all comes out in the wash. Who's going to stop us? If the private ice cream industry can't generate enough smiles, then it's time for democracy to grab the scoop.*

In this way, a public investment authority could program the market to start maximizing whatever it is we democratically direct it to maximize, and the question of what kind of life is worth living reveals itself to be a social rather than an individual one. Omarova has a good list of concrete collective tasks worth accomplishing, and decarbonization is at the top. In this proposal, the green subsidies aren't like a juicy steak tossed to a hunting dog; they're like a big wooden horse rolled onto the capitalist doorstep that the grabby oligarchs can't help but accept. With the right initial appropriation, an institution like the NIA could do more than marketcraft the energy transition: It could *socialize the means of investment.*

GREEN STICKS

Socializing the means of investment could theoretically allow the American polity to revalue its values and compel capitalists to dance to a new tune. It reminds me of the 2001 kids' movie *Monsters, Inc.*, which concludes with monster society switching its mode of energy generation from children's screams to laughter. But it's not the revelation of an alternative monster-fuel source alone that transforms the social metabolism between the two worlds; the heroes still have to repress existing forces that are not just wedded to scream extraction but also invested in

* Ice cream smiles are decarbonization in this metaphor, to be clear. We could also *literally* finance ice cream according to scoops if we wanted to, though. I say we try it.

updating the technology. (The two villains, who represent an alliance between scream capital and scream-harvesting labor, are arrested and exiled, respectively.) As I recount in this book's introduction, Shell has planned to take advantage of the green energy boom *and* profit from fossil fuels as long as it can. The point of democracy offering capital a bundle of green carrots is to actually build a decarbonized energy system, not to cover financial bets for the oil and gas industry. Alongside the marketcraft hand that giveth, we need one that taketh away.

Marketcraft has plenty of unpleasant tools as far as capitalists are concerned. If we as a democratic collective are engineering the economy to serve our shared needs—including and in particular decarbonization and climate-change adaptation—then that means making choices, positive and negative. The supply-side crafters emphasize subsidies and other forms of spending, hoping they can crowd out socially irresponsible investment without starting too many fights with the fossil fuel industry's representatives. But there are advantages to employing what will appear to at least some parts of capital as *dark marketcraft*. It might keep investors from trying to have their cake and eat it too, as well as speed us toward our actual social goals. Democratic coercion teaches all of society an important lesson about who's really in charge. If the people were the boss, and the whole world were really at stake, would we let centimillionaires and billionaires bop around on private jets?

And yet the number of global private flights was at an all-time high in 2022—5.4 million.[26] Of these, an 85 percent supermajority are private flights *within* the United States, jaunts of climate-destroying, carbon-intensive convenience taken by the American 1 percent, making up around one-sixth of total domestic air travel.[27] But even if the numbers weren't dramatic, and even if air travel weren't a stubbornly unelectrifiable source of fossil fuel consumption, I think there'd be something worthwhile about challenging private jet usage just on principle. A marketcrafting strategy has to maintain credibility with the public, and that means the state has to show itself willing and able to protect public interests from its partners in the capitalist class. Private jet owners are a perfect target.

The simplest negative way to craft a market is a sumptuary tax, designed to make forms of excessive consumption suppressively expensive. Researchers Omar Ocampo and Kalena Thomhave suggest taxes on "short-haul flights, jet fuels, and sales of both new and pre-owned aircraft" that would affect the 1 percent of travelers responsible for half of all aviation emissions.[28] Such a levy would be less about the revenue—marketcrafters are comfortable with the idea that the government gets money by printing it, not by raising taxes—than about discouraging rich people from choosing to take private flights. But private flights are already quite expensive. Forcing the rich to pay the actual cost would be a defeat for the industry's lobbyists, yet it's not at all clear that these taxes would be prohibitive. Besides, executives could continue to pass the costs along to their shareholders, private travel having long been established as a top-tier perk in corporate America.

If we really wanted rich people to have to wait around at the gate with the rest of us slobs—or even warm the pews at the train station—we'd shut off their access to the public goods that make private air travel possible: the runways, the air traffic controllers, the air routes themselves. The Europeans are already getting started. Two Dutch airports have advanced plans to end support for private flights. France has banned domestic flights to destinations less than a two-and-a-half-hour train ride away, ending routes between Paris and the cities of Bordeaux, Lyon, and Nantes. Spain plans to follow suit. The EU has so far declined to take policy action, but there's increasing noise from member states as climate activists target airports and demand an end to the practice.[29] Given how US-centric the industry is, one American legislative index finger across the throat could decimate private air travel.

In addition to a moral victory for the working class and an emissions reduction for the planet, such an action could earn the United States some much-needed climate credibility around the world.

Taxes and bans can craft capital markets, too, corralling investment into socially useful areas. For example, the journalist Harold Meyerson has pointed to the small supporting role that the Biden administration's 1 percent tax on buybacks—the practice of firms spending cash to boost their stock prices by buying shares off the market—plays in the green investment agenda.[30] Meyerson's BUYBACKS ARE DOWN, PRODUCTION IS UP is exactly the kind of headline the administration sought out with its green industrial policy. Biden called for increasing the tax to 4 percent, and a group of congressional progressives has proposed an all-out ban on the practice.[31] No doubt corporate managers can find other ways to disgorge capital to their shareholders, but grading the terrain against it shapes their decision-making process. If capitalists are always looking for an easy way to make money, democracy should exercise its prerogative to slam some doors.

One of those easy ways—of which capital has made extensive use over the past decades—is monopolization. Rather than competing, many of the twenty-first century's most celebrated firms have assembled huge war chests and bullied their competitors out of the arena with predatory techniques—that is, when they're not buying or merging with their competitors outright. Antitrust enforcers have long relied on a consumer-welfare standard, one that holds that as long as Amazon and Google push down prices, they can't be all bad. But a new cohort of regulators wants to use a higher bar.

Advocates for crafting a more competitive market were cheered when Biden named scholar and public intellectual Lina Khan to head the Federal Trade Commission in 2021. In her introductory letter to the FTC staff, she revalued the agency's values, pointing to workers and small businesses, not just price-conscious consumers, as parties with an interest in strong antitrust laws. In her critique of private equity, Khan called out investors for noncompetitive behavior and consumer-protection violations as well as for distorting ordinary incentives "in ways that strip productive capacity."[32] This suggests that Khan's intentions go beyond refereeing the market in accordance with some transcendent idea of fair play, and venture into making sure that the nation's labs and factories are building the actual things we need. This marks a philosophical turn from market fundamentalism toward administrative marketcraft. Khan has stirred up the FTC with her confrontational "litigate, don't negotiate" policy, picking fights with big tech firms such as Facebook and Microsoft. But perhaps her biggest intervention was in the labor market, where she has attempted to ban noncompete agreements for workers. Though her win percentage isn't very high at the time of this writing, Khan's aggressive FTC is already crafting the market merely by flashing its teeth.*

When it comes to marketcraft, environmental restrictions sometimes get a bad rap, especially if they tend to obstruct the construction of green energy installations. But straight-up environmental rules aren't just about protecting the earth directly; they craft markets as well. When fossil fuel firms are weakened by regulatory marketcraft, that leaves them vulnerable to green competitors and other political actors. After the institution of Obama-era EPA rules on coal emissions, for example, the nonprofit Sierra Club's Beyond Coal initiative pursued a lawfare strategy to block and decommission as many plants as possible. Buoyed by tens of millions of dollars from philanthropist and clean-energy investor Michael Bloomberg, Beyond Coal has exploited the coal industry's rising compliance burden by petitioning the boards that oversee the country's public-private electricity systems. Writing for

* One way Khan has shaped the market is by influencing a generation of law students. See Marcia Brown, "The Next Generation of Law Students Is Obsessed with Lina Khan," *Politico,* November 6, 2023.

Politico, journalist Michael Grunwald explains that "while the utilities might be happy to charge their customers tens of millions of dollars for upgrades in order to comply with one new rule—plus a tidy profit they're usually guaranteed for capital improvements—utility commissions might not let them start down that road if they faced hundreds of millions of dollars in additional compliance costs from rules still to come."[33] Those were good bets, and the Inflation Reduction Act has indeed accelerated the coal drawdown; one analysis has the industry declining to 37 percent of its 2011 peak capacity by 2030.[34]

GREENCETERA

It's easy to fall into the trap of thinking about marketcraft solely in terms of the government shaping the competition between market actors. Whether declaring "Up for grabs!" with billions in subsidies or stapling an OUT OF BUSINESS sign on a coal plant that's been regulated to death, crafters put the state's thumb on the scale, picking winners and losers. Those are big parts of the strategy, but there are lots of policy actions that don't fall cleanly into either the carrot or stick category. We take the stuff in this third basket for granted, by design, partly because it *is* granted and partly because the details can be very complicated, even boring. Consider highways, for example, which the trucking industry can assume will still be there tomorrow, next week, and next year—even though that requires many, many specialists working behind the scenes with variables we might not understand. Firms depend on the public education system to teach their workers how to read, and school bus manufacturers depend on the same system to buy buses. Government-funded labs at public universities produce all sorts of research that winds up in industry hands. The state issues permits that confirm oversight of private construction projects. Guiding the market into green transition means being attentive to the many ways public policy is the stage where market actors play their parts.

Decarbonizing the transportation and electricity sectors quickly—the marketcraft strategy's central purpose and justification, recall—requires a lot of public rule making and coordination. With the Biden administration's surprising success wheedling so much spending out of Congress, it's this third category of action that often seems to be the holdup. What's the point of building electric vehicles if buyers don't have convenient places to charge them? And what good is a solar or wind farm if there's no adequate transmission line to connect it to the electricity grid? Insufficient marketcraft in these areas creates choke points for decarbonization and, with them, dangerous delay.

Market-fundamentalist legacies have left the United States trailing when it comes to implementing standards for green infrastructure. It's gotten so bad that even management consultants (not known for their social-mindedness) are begging the state to step in. Note the conclusions of a 2021 Boston Consulting Group report titled "How Governments Can Solve the EV Charging Dilemma": "A well-governed ecosystem can help the market avoid the competing standards, redundancy, and lack of interoperability found in many rollouts to date. The governing body should therefore act as a central coordination and orchestration unit for the EV charging ecosystem," the authors write, with characteristic consultant aplomb. "In fact, we see a centrally coordinated market as a key success factor, with e-mobility targets, consistent action plans, and central economic support."[35] Where do these capitalist guns for hire see that success happening? Norway, the Netherlands, and China. Not the traditional models for American policymakers.

Rather than plan a national EV charging network, the United States let the anarchic market try to figure it out. The result has been a clusterfuck of nonstandard standards, as American roads host cars with three different types of plugs: the Combined Charging System (CCS, begun in Europe), CHAdeMO (out of Japan), and Tesla's own North American Charging Standard (NACS, which is, at the time of this writing, confusingly, not yet a standard).[36] Imagine if gas-powered cars had three different nozzle couplings and you had to remember to always go to the right station. And because motor vehicles have proprietary apps now, any given charging point might not let you pay to use it, even if you have an adapter for the plug. The EU's common CCS standard hasn't made the single market a seamless territory, either, because some charging points force users to download apps or join subscription programs. China, by comparison, has the single GB/T charging-plug standard—Tesla has to use it there, too—and drivers pay by QR code via the ubiquitous WeChat and Alipay apps.[37]

After a decade-plus of failed American laissez-faire, the Biden administration embraced its role as EV mediator. The Bipartisan Infrastructure Law created a new Joint Office of Energy and Transportation to coordinate between the Departments of Energy and Transportation as well as industry, in particular to figure out the charging infrastructure, including $5 billion in state grants for charging points.[38] One of the first things the joint office did, after Ford and GM announced they planned to adopt the Tesla plug, was to get Tesla to partner with SAE International (formerly the Society of Automotive Engineers) on an expedited standardization of the NACS connector.[39] Here, public policy and engineering slide together smoothly without the petroleum-based lubricant of finance.

The joint office also established the National Charging Experience (ChargeX) Consortium, bringing together the Department of Energy's national laboratories with dozens of industry stakeholders. This is the kind of collaborative planning needed to produce the infrastructure drivers will take for granted, such as a common protocol for charging points to use when transmitting live price, location, and availability data—or, more broadly, a national EV charging network that *just works*.[40]

That's marketcraft behind the scenes, and the bureaucrats at the tiny joint office are some of the green transition's uncelebrated prop masters, making sure everything works and is in its right place, setting the stage. Among their mandates is one to help school districts across the country access their share of $5 billion in EPA appropriations for electric school buses, most of which has now been deployed.[41] Without that kind of unglamorous coordinating action, billions of dollars in spending threaten to become mere words on a page.

The electrical grid's coordination problems could fill a book twice the size of this one. At the end of the twentieth century, in a perfect example of how marketcraft defies the regulation/deregulation binary, the Federal Energy Regulatory Commission (FERC) put tons of rule-making effort into creating more dynamic and commercial electricity markets. The result is a strange mosaic of public-private coordination in which some states and regions stick to vertically integrated, monopolistic utilities handling generation and transmission while others operate under regional transmission organizations that assemble competitive wholesale generation markets, which may or may not overlap with an open retail market that allows consumers to choose among private suppliers.* The FERC's idea was to invite private capital in to update and improve the grid, but as we've seen, it takes more than an invitation to craft a good market.

The most frequent complaints you'll hear about the electricity market on the supply side are "permitting" and "the interconnection queue." As a general whine about government interference, "permitting" works great because it could apply to any number of things, from a federal environmental impact report to local construction approval.

* For a concise overview of US electricity regulatory schemes, view the EPA's summary at epa.gov/green-power-markets/us-electricity-grid-markets.

Fossil fuel companies can join in the unhappy chorus and even lead it, demanding a fast track for their pipelines through the federal bureaucracy. That is not the same as a solar farm that runs into a land-use restriction, but it comes to seem that way. The result is a substantial amount of pressure on the government at all levels to "get out of the way," even though we know perfectly well that the same industry relies on the state to *make* way in the first place. The situation calls for more specificity.

In the world of marketcraft, answers often come in the form of reports, and the permitting controversy is not an exception. Researchers at the Climate and Community Project wrote a 2023 report offering a "progressive take on permitting reform" that brings much-needed clarity to the subject.[42] It's key, the researchers write, not to allow concern about the rapid deployment of clean energy infrastructure to become an excuse for fossil fuel pipelines to evade environmental regulations. When they looked at whether NEPA—the big bad permitting bogeyman—was really blocking clean energy projects, they found that the vast majority get exclusions from the most onerous and time-intensive requirements because the feds know that a moderate-size solar farm, say, does not usually pose a large, complicated environmental hazard. Insofar as there's a problem with speed, it's something that can be solved with more staff. Rather than cutting red tape, the report concludes, the federal government should wrap said tape around the fossil fuel industry's nose and mouth, denying *all* permits for new oil, gas, and coal projects. Killing the competition is the perhaps unexpected but perfectly logical way that permitting reform could most quickly improve the clean energy industry's fortunes.

As for the market flaws that are truly delaying the deployment of a green grid, they don't have anything to do with too much regulation. In fact, the regions and states that stuck to integrated utilities have had an easier time committing to big investments in new nuclear power plants, for example, because they can count on their customers to pay for it. Much as in the electric vehicle charging-network case, America doesn't suffer from a lack of marketcraft so much as an abundance of *bad* marketcraft. The Climate and Community researchers conclude, "Environmental review is often blamed as a cause of delay in infrastructure development, but long interconnection queues to connect to the electricity grid are the main cause for delay with renewable energy projects."[43] When you build a commercial-scale energy-generation and/or storage facility, you can't just plug it into a wall socket and start feeding juice back onto the grid. The transmission operator first checks if the project is feasible, then studies what kind of impact it will have on the existing grid, then decides where the connection should go and what kind of equipment and upgrades are necessary and who should be on the hook for what percentage of the costs, and then they're in a place to negotiate a contract, which may or may not be acceptable to the developers.* This process can take years, and as a result, developers have flooded the queue with speculative projects, the majority of which they withdraw at some point, which is itself a significant cause of delay, as is a lack of the long-distance high-capacity transmission lines needed to connect site-dependent solar and wind resources. The FERC issued some new rules in 2023 calling for transmission operators to plan better and switch from first-come to first-ready, along with financial penalties for both sides if they fail to meet their obligations in a timely fashion.[44]

It is not a bad idea for operators to "conduct regional transmission planning on a sufficiently long-term, forward-looking basis to meet transmission needs driven by changes in the resource mix and demand," as the FERC has instructed them to do.[45] But there's a basic mismatch between the ask and the asked. As Emma Penrod reported

* See Joseph Rand et al., "Queued Up: 2024 Edition—Characteristics of Power Plants Seeking Transmission Interconnection as of the End of 2023," Lawrence Berkeley National Laboratory, April 2023.

in *Utility Dive,* "Permitting is still managed at the state level, where each state sets its own rules and priorities."[46] The offices responsible for interconnection are clearly understaffed, yet it's not clear to me that the problem isn't in the system's structure itself. With so many different authorities operating across the public-private divide, coordination becomes intrinsically difficult. We should be planning clean energy infrastructure at the continental level, but instead we have the Maine Public Utilities Commission issuing its own interconnection rules and procedures.[47] A radically progressive marketcrafting strategy, it seems to me, would look back to the 1990s and conclude that the current electrical-generation and transmission market was (a) made, (b) made poorly, and (c) could be remade better.

Electric vehicle charging networks and electricity transmission systems are weird, important markets, and the only way they're going to be tolerable for consumers is if the state coordinates them. Even the capitalists admit as much. Hewing to false binaries such as regulation versus deregulation distracts us as a society from the work of planning and building the things we need. Marketcraft as a strategy busts the binary and asks the public—through our representatives but also directly, as a mass of democratic subjects—to take responsibility for the outcomes our systems produce. Policy is about choices: As far as marketcraft is concerned, blaming capitalists for the slow interconnection queue is like blaming the stuffed animals at your imaginary tea party for the poor quality of the cucumber sandwiches.

I've focused on decarbonizing transportation and the electrical grid because those areas are what marketcrafters tend to identify as the most important objects for their strategy in the near term, but there are many related areas of public policy where an active crafting approach could generate improvements for people and the climate. California's Right to Repair Act will compel electronics manufacturers to make high-end tech parts, tools, documentation, and software available for seven years, transforming the global market and thwarting the highly wasteful manufacturing strategy of planned obsolescence.[48] Spurred by a COVID lockdown surge, state legislatures have made it easier for so-called cottage food producers to package and sell their homemade goods. At the local level, cities have opened public spaces to farmers markets, and the USDA chipped in $75 million for the Local Meat Capacity Grant Program—all parts of building a more resilient, agro-ecological food system.[49] There's a lot of work for governments to do in crafting the labor market for a green transition, whether by founding a national university to provide young people with a free education in the many climate-related subjects in which we need experts or by creating an Office of Green Jobs that connects workers to climate-focused openings in firms that have agreed to certain labor protections. Researchers at the Political Economy Research Institute have even called for a public credit-rating agency in order to seize the means of asset evaluation from the private clutches of S&P, Moody's, and Fitch.[50]

A society in which the radical marketcrafters win might have a lot of the same stuff as our world. There would still be jobs that pay us money and stores where we could buy things we need and electrical outlets where we could plug them in. But the kinds of lives worth living would be different. The combination of suppressive regulation and superabundant carbon-free electricity would overwhelm the fossil fuel industry—perhaps transforming it into the carbon storage industry, perhaps dissolving it altogether. The value of oil falls to a minimum, and it ceases to be worth the trouble. Capitalists still exist, but their ability to direct society is significantly curtailed: Like public employees, they're made aware that their contingent license to operate comes from the rest of us, and we require them to be useful, not merely profitable. As mute economic coercion loses its bite, we're increasingly free to plan our society according to human reason—whether we're sharing a surplus or coping with a disaster, both of which we'll certainly confront. This is no utopia; it's here, made better, by us and for us.

Now: challenges to the marketcraft strategy.

My purpose in exploring the challenges to each of the strategies under consideration is not to compare and contrast in order to find the right answer; I expect people are going to be working effectively under and across each of these strategies. With that in mind, I ask the reader to consider the objections not as claims about errors or holes. The point of this book is not to find a plan without flaws but to map the jagged coastlines of the real, flawed plans we have in the hope that we might be able to fit them more auspiciously into the currently elapsing reality. For that purpose, it's fortuitous that none of the strategies can claim to hold all the solutions within its neat boundary.

II.

US AGAINST THE WORLD

As I wrote in the introduction, climate change and capitalism are both decidedly *global* problems, yet the discussion of strategy so far has focused almost exclusively on the relations between the US government and domestic capital. This is (I promise) not the result of my drifting attention: It reflects a deep and genuine ambiguity within not just climate discourse but also climate policy. For many people around the world, the response to Western green politics as a whole might be a more earnest version of what Tonto told the Lone Ranger when the pair found themselves surrounded by hostile Indians: "Who's this 'we,' white man?" National marketcraft does not always have a great answer, and there is no such thing as decarbonization in one country.

In the English language, few critics have tackled this issue as well as Max Ajl, whose 2021 book, *A People's Green New Deal,* refuses to shy away from the global nature of the questions at hand. "Much of left-liberal climate talk is based on administering rather than eliminating capitalism," he writes, accurately, "and as a result is built on a seldom acknowledged foundation of assumptions regarding the global distribution of wealth and consumption, and the institutions with which it is tied, in terms of why emissions are produced and their consequences, which are intimately related to which lives matter and which lives do not."[51] What good is a dense electric vehicle charging network in Poughkeepsie to Grigris, the Chadian dancer turned petrol smuggler? Freezing the consumption of everyone in the world at current levels preserves global inequality, leaving some people cruising in EVs and some people burning wood as cooking fuel.

If all future carbon emissions were fully mitigated tomorrow, as Ajl notes, the world would still suffer the consequences of accumulated historical CO_2, which has led to the extreme and worsening weather we are already experiencing. Those consequences tend to fall hardest on poor people and poor countries, both because there's been little investment in climate adaptation in those places and because environmental fragility is a consequence of colonialism and postcolonial exploitation—the environmentally unequal exchange in which the global South's topsoil, for example, gets shipped to the North in the form of plantation-monocropped tropical fruit. There is a desperate need for finance throughout the world to fund planned decarbonization as well as adaptation for the wholly predictable disasters that are already hurtling toward us through time... *and* general improvements in living standards. But whereas America can deficit-spend its way to a green grid, "the pandemic, commodity price shocks, and US interest-rate hikes have left global South countries with limited fiscal room to invest in cutting emissions and building resilience against climate change," one analyst concluded.[52] And yet Western financiers and the policymakers who craft their markets have not been spurred into compensatory action. A report from the Climate Policy Initiative on global flows of climate finance between 2011 and 2020 found that "75% of all climate finance was concentrated in North America, Western Europe, and East Asia & Pacific (primarily led by China)."[53] The communities around the world that are most susceptible to climate disaster are also those with the least access to the financing they need to adapt, and many, many people will perish as a result. And that's if the American marketcraft strategy goes exactly according to plan.

The United States and the EU have already adopted fortress mentalities with regard to climate refugees, tightening borders and making deals with other nations, including Mexico and Turkey, respectively, to keep asylum seekers at bay. It seems clear that insofar as marketcrafters have gotten hold of the climate-policy steering wheel, they've stayed within their home waters. Transnational capital's caution about infrastructure and adaptation investment in the global South becomes a self-fulfilling prophecy: Who wants to sink money into a disaster zone waiting to happen? To change the situation, Western investors are going to require a huge risk premium, and no one seems both able and inclined to provide it.

Forget solidarity with the Third World; green marketcraft is making it hard to hold the First World together. Economist Daniela Gabor has critiqued the dominant "derisking" strategy that lowers private-investment hurdle rates with bribes to capital—the strategy's positive, subsidizing aspect.[54] Pushed by American support for EV manufacturers, she writes, European priorities have shifted from "the EU wanting to decarbonize quickly, even with Chinese clean tech imports," to "global market share in clean tech."[55] The French president, Emmanuel Macron, has called for a Buy European Act, complaining that "you have China that is protecting its industry, the U.S. that is protecting its industry and Europe that is an open house."[56] However, certain countries within the EU are in a better place than others to derisk their national industries, prompting fears that Germany and the Netherlands, for example, will be set to delink their fortunes from the larger union, leaving smaller markets behind. A transition based on national marketcraft strategies pits countries against one another in a subsidizing race to the bottom, courting mobile capital across borders and tilting the strategy's benefits toward bosses and away from workers. If the country that's most competitive in capitalist terms wins, the people of the world are going to find themselves worse off at the finish line.

American subsidies for domestic electric vehicles under the IRA have proved controversial with other allies as well, in particular the ones who build cars. South Korea threatened to join the EU in a World Trade Organization complaint after Hyundai (including its subsidiary Kia) found itself unexpectedly shut out of the EV tax credit program, prompting "emotional and political repercussions," according to the country's trade minister.[57] The South Korean government committed to its own set of domestic EV tax credits, suggesting that a green trade war wasn't far off. The IRA had already begun luring foreign—especially South Korean—capital into direct investment in the United States for EV and battery factories, often intertwined with domestic producers. But sourcing restrictions threatened to lock out a number of friendly countries—such as South Korea as well as Japan and the EU—that did not meet the law's "free trade agreement" threshold. Stuck between the rock of Joe Manchin's protectionism and the hard place of a trade war among friends, the Biden administration used marketcraft magic to try to set things right with the allies, instituting some new critical-minerals free trade agreements that sneak countries under the velvet rope with a clarification that *leased* electric vehicles from abroad are tax-credit eligible. This move infuriated Manchin, and he symbolically pulled his support for the bill, but he had already voted for it, and the IRA is law. Still, these moves undermined parts of the bill. As one analyst concluded, the leasing clarification "dulls the incentives for firms to shift their battery input supply chains out of China."[58] Those incentives are the only thing standing between the domestic battery industry and Solyndra's fate.

The third and in some ways most important international conflict that the American nationalist marketcraft strategy generates is with the

People's Republic of China, but this is one fight the Chinese started. After the 2008 global financial crisis, with a left-wing faction within the Chinese Communist Party pushing for greater state involvement in the economy, the CCP adopted an aggressive marketcraft strategy, pumping huge amounts of money into selected sectors, including aluminum, solar panels, and EV batteries.* China had been the "workshop of the world" since its accession to the WTO, at the end of 2001, but in these sectors, it wouldn't be waiting for instructions from foreign multinationals. In EVs, China's sizable state fleets led the charge, contracting for initial orders. On top of nearly $30 billion in subsidies and tax breaks granted to producers between 2009 and 2022, EVs were exempted from a consumer license-plate rationing scheme meant to reduce air pollution.[59] The Chinese EV battery winner, CATL, quickly became the world leader, not just in terms of scale but also in terms of advanced research and development, capturing more than a third of global market share at the time of this writing and helping China dominate the production of battery cell components—including more than 80 percent of anodes.[60] Even Ford is licensing CATL's technology for its IRA-compliant Michigan battery factory, much to the chagrin of Sinophobic Republicans.

In 2017, low prices and sophisticated supply lines helped lure Tesla, the world EV leader, to Shanghai, where it built its third "gigafactory"; the firm is set to build a battery-pack plant there as well. The existing factory is extremely productive, and Tesla exports the cars to Australia, Europe, and, as of the spring of 2023, to Canada.[61] The IRA's restrictions have kept it out of the United States so far, but the government is then effectively paying people not to buy Chinese cars, which is not exactly a climate plan, never mind a *good* climate plan.

* In marketcraft, too much is often just right, and subsidized low-range first-generation Chinese EVs now sit abandoned and unused by the hundreds in urban "graveyards." See Linda Lew, Chunying Zhang, and Dan Murtaugh, "China's Abandoned, Obsolete Electric Cars Are Piling Up in Cities," Bloomberg, August 17, 2023.

And yet it has become almost compulsory: If the US government seeks to craft an EV market—and an advanced solar-panel market, and a wind-turbine market, and a green aluminum market, and so on—that will attract private capital, then it needs a plan to make those industries investable, which means making sure they won't drown in subsidized (and therefore cheap) Chinese imports, the way Solyndra's solar cylinders did.

Instead of the two largest economies coordinating to transition the world to renewable electricity, the vagaries of global capitalism have them competing, as if we don't share a single atmosphere. In 2018, Donald Trump imposed tariffs and trade restrictions on China, and not only did Joe Biden keep them in place, his administration also issued surprising export restrictions on high-tech materials, a naked attempt to slow China's economic rise.[62] And while solar tariffs against China have led to some expansion of the puny American solar assembly industry, the much larger installation industry has suffered, and with it so has the marketcraft strategy for decarbonization.[63] The trade war shows few signs of declining on its own: With its annual foreign direct investment into the United States falling from $46.5 billion to only $5.4 billion (as a consequence, in part, of the Trump tariffs), Chinese capital is increasingly global, investing in infrastructure projects around the world under the Belt and Road Initiative, expanding extraction and manufacturing into neighboring countries, forging corporate-led production deals with strategic economic partners such as Turkey and India, and exporting many millions of electric cars and solar panels everywhere but the United States.[64] Meanwhile, American tariffs on offshore Chinese solar-panel producers located in Thailand, Vietnam, Malaysia, and Cambodia resumed in June of 2024.[65] If we're not careful, nationalist marketcraft will drag this warming world into another cold war.

There is another option: *Inter*nationalist marketcraft. In their book *Fixing the Climate: Strategies for an Uncertain World,* Charles Sabel and David Victor look for a precedent-setting example of effective global cooperation on the environment and find a good one in the Montreal Protocol, the late-'80s agreement that set up a framework to halt the use of ozone-depleting substances (ODS). It was American-led, but that's because the country had instituted the first ban on nonessential ODS in the late '70s and had substantial credibility on the issue. The Montreal framework was scientist- rather than politician-controlled, with a Scientific Assessment Panel operating alongside a Technology and Economic Assessment Panel, the latter of which oversaw a number of industry-specific Technical Options Committees (TOCs)—refrigerants, foams, solvents, aerosols, and so on. As the TOCs came to consensus on new deployable alternative technologies, they would revoke the exemptions that allowed, for example, medical inhaler manufacturers to keep using ODS. "For the scientists and engineers organizing the new regime," Sabel and Victor write, "there was nothing novel about the idea of collaboration in the solution of shared problems. As they were connected by education and work experience to professional communities that value reciprocity and honor elegant solutions, such cooperation was almost a reflex."[66] This global economic planning worked—the ozone layer turned around, and recent modeling published in *Nature* suggests that Montreal may have saved us from a worst-case global warming situation in which UV radiation damages plants' ability to absorb carbon and catapults us past six degrees of warming by the twenty-first century's end.[67] Yet rather than apply the world's collective wisdom and cunning to the world's collective problems, capitalist competition makes it hard for countries to even approach these questions in a reasonable way at the domestic level.

GODDAMN CARS

If there is a commodity into which the marketcrafting transition strategy crystallizes—and there is—it's the electric vehicle. While the IRA's tax credits for energy generation are agnostic, the extensive support for EVs is specific. Despite creating all sorts of international problems, cars are the crux of the American climate strategy. On the surface it makes sense: Internal combustion engines are the cause of the lion's share of direct American household fossil fuel consumption, especially as the grid decarbonizes. But as I've said, it's *in*direct consumption of fossil fuels—through agricultural products, plastics, and basically everything we consume, period—that drives emissions. It's easier to change the cars than it is to change where people are going. If the marketcraft strategy is a compromise between a reasonable democratic approach to the climate crisis and the demands of capital, how much reason do we have to sacrifice to the status quo? And how can we be sure it's not too much? Here's one indicator I think we can use: The number of cars should go down, not up.

A reasonable society would not have nearly a car for every person. That America does is a sign that we live in an unreasonable society. We as a public subsidize cars in innumerable ways, including by ceding much of public space to their coming, going, and parking.* Much as they did with cigarettes, researchers are continually discovering new ways that cars are bad for society: As I write, there is new attention on toxic pollution from tires, pollution that has turned rainwater runoff deadly for spawning coho salmon on the West Coast.[68] And then there's the climate crisis: In 2023, a team of physicists from Imperial College London concluded, in admirably plain language: "[A]s well as implementation of emission-reducing changes in vehicle design, a rapid and large-scale reduction in car use is necessary to meet stringent

* See Donald Shoup, *The High Cost of Free Parking* (Routledge, 2005).

carbon budgets and avoid high energy demand."[69] Building a lot more cars, it suffices to say, will not help with that.

The United States has long been egregiously car-dependent, and faced with a global catastrophe that absolutely requires the decommissioning of basically every single car as soon as possible, the prevailing plan seems to be to replace the fuel source and do the same thing all over again, at tremendous cost. An influential study on green transition requirements from Princeton University suggests that there will be a need for 328 million electric cars in the United States by 2050.[70] That is not a good sign for the climate transition's reason ratio.

That pessimistic conclusion about proportions is more or less the same one as the analysts at PwC (formerly PricewaterhouseCoopers) came to when they looked at venture capital investments in the climate sector. In their 2022 annual report—the third of its type—the headline news wasn't good (there is "relative stability in venture capital investment at a moment when sharp increases are needed to meet emissions objectives"), but the conclusions at the bottom were worse. Climate VC investment is overwhelmingly concentrated in areas where the analysts rate the emissions-reduction potential as low, in particular "light-duty battery EVs," which gobble up the pie's largest slice by far. The mobility and energy categories captured 75 percent of climate venture capital between the third quarter of 2021 and the third quarter of 2022, though those categories are only responsible for 27 percent of emissions.[71] Food, agriculture, land use, industry, and the built environment as categories all suffer from underinvestment relative to their emissions. VC money is also concentrated in more mature technologies—solar panels, electric scooters, and, of course, cars—rather than in the less mature categories that need adventurous capitalists to back them, such as food-waste reduction and carbon capture. All in all, it's enough for PwC to come to the damning conclusion that venture capital is "an inefficient market for investing in climate outcomes."[72]

The larger financial markets aren't as yet any better focused than the VCs are. A meta-analysis from the Climate Policy Initiative found that, while estimates for needed annual climate-related investment in global agrifood systems range from $212 billion to $1.267 *trillion,* they tracked the actual investment at only $28.5 billion.[73] The energy transition has lured asset managers deeper into the climate-related energy, transportation, and water sectors, where private-equity deals jumped from around $60 billion in 2020 to nearly $150 billion in 2022.[74] But that doesn't necessarily bode well: As Brett Christophers points out in his book *Our Lives in Their Portfolios: Why Asset Managers Own the World,* real asset funds would rather buy stuff that already exists, wrench more money out of it, and sell it on. They much prefer it to building new infrastructure, which is the plan in less than 20 percent of these investments.[75] "What the evidence both before our eyes and in fund performance data shows," he writes, "is that actually holding the asset—let alone stewarding it—is really not what the business is about."[76] For financial actors, "exposing" themselves to the green transition's upside is not the same thing as building and maintaining it.* A number of electric car manufacturers, however, have raised significant capital.

I have a suspicion that part of the reason cars play such a big role in the imagined marketcraft transition is their proponents' fondness for World War II analogies that conflate the war-materials production effort with the postwar "golden" age of car-centric capitalism. State-led planning run largely through private companies quickly built the arsenal needed to win the war, then the companies turned around and built a stronger, more inclusive capitalist economy around high-wage

* For more, see Adrienne Buller, *The Value of a Whale: On the Illusions of Green Capitalism* (Manchester University Press, 2022). On the structural challenges of definancialization, see Sahil Jai Dutta, "Countering Financial Claims," in *Capital Claims: Power and Global Finance,* ed. Benjamin Braun and Kai Koddenbrock (Routledge, 2023), 50–68.

production jobs, such as manufacturing cars. A well-crafted green transition should do both at the same time, and EVs seem like just the object: They're weapons in the war on global warming, generators of manufacturing jobs, and novel consumer commodities all in one. But new research suggests that our story about the twentieth century is too simple and that there are as many lessons there about what *not* to do as the opposite.

Economic historian Alexander J. Field spent half a decade researching his book *The Economic Consequences of U.S. Mobilization for the Second World War,* which argues convincingly that the war-materials production effort did not go nearly as smoothly as we like to remember, nor did it have a salutary effect on the postwar economy. Instead, what his research shows is that the oil industry used its influential position to sabotage the national planning effort in its own interest and set itself up to profit afterward. It's not an overstatement to say that allowing American fossil fuel executives to participate in that coordination nearly lost the Allies the war.

Field's signature example is the US synthetic rubber program. Though historians traditionally consider the program to have been an industrial miracle, he finds a series of preventable errors. First, cheap rubber imports from Southeast Asia made US firms unwilling to hold significant reserve stocks—*What if the price goes down?!*—or invest in planting large amounts of guayule, a shrub indigenous to the American Southwest from which it was proposed to extract significant amounts of rubber.* This left the United States so underprepared when Japanese advances cut off rubber supplies that the country was forced to restrict automobile travel not for lack of gas but for fear of running out

* Field notes that government scientists at the Agriculture Department repressed wartime guayule research in part because the most advanced researchers were a team of Japanese American agronomists working from the Manzanar concentration camp. Alexander J. Field, *The Economic Consequences of U.S. Mobilization for the Second World War* (Yale University Press, 2022), 86–87.

of tires. Industry dragged its feet, refusing to invest in synthetic capacity, since it assumed the cheap natural rubber would start flowing again after hostilities concluded—an attitude that ignored the fact that the United States could *lose the war*. When the synthetic program did get up and running with government money, the oil industry insisted on using petroleum (rather than easily produced alcohol) as a feedstock, unnecessarily slowing output at a crucial time. Field concludes that the wait for synthetic rubber delayed the American invasion of France on D-Day by a year.[77] "The decision to structure the program around an almost exclusive emphasis on petroleum as a feedstock," he writes, "worked at cross-purposes with the immediate objective of winning (or at least not losing) the war."[78] We can imagine a future historian saying something similar about today, trying to answer the perplexing question of why, at a historical turning point for the planet, humanity spent such a large percentage of its collective resources on electrifying Americans' cars: The decision to structure the transition program around electric vehicles worked at cross-purposes with the immediate objective of cooling (or at least not heating) the world.

Underlying the marketcraft strategy for a green transition is the sometimes unspoken, sometimes overt idea that capital will use substitute technologies to *replace* fossil fuels. But this attitude begs the question of whether replacing what Americans have today at the commodity level is the right strategy at all, and it assumes that substitution is a natural cure for odious forms of production. Here, too, the historical evidence isn't as supportive as popular recollection suggests. Perhaps the proudest social-justice achievement in capitalist history is the way British free-trade advocates helped shatter the English slave industry. Representing the rising industrialist fragment of the ruling class, they pitted their commercial interests against monopoly deals that supported the colonial slavers of the West Indies. This was the rise of liberalism and its free market, breaking the literal chains of servitude in the course of the supposedly selfish pursuit of profit. But as Eric Williams points out in his history of British abolition, *Capitalism and Slavery,* "British capitalism had destroyed West Indian slavery, but it continued to thrive on Brazilian, Cuban, and American slavery."[79] British importers of American cotton and Cuban sugar opposed the political bloc of Anglo-Caribbean slavers but not slavery itself. Market substitution did not end slavery—that took insurrection and war.

We can see something analogous in the relocation of emissions. If an asset fund divests from a coal plant and invests in a solar farm, it doesn't mean the fund managers are "against" coal: It just means the coal they're relying on is in China, where it's used to power polysilicon wafer factories in Xinjiang. "For years, China's low-cost, coal-fired electricity has given the country's solar-panel manufacturers a competitive advantage," the *Wall Street Journal* reported, "allowing them to dominate global markets."[80] The calculation is that even the dirty production of solar panels is worth it in carbon terms over the panels' life span, but the data that calculation is based on is questionable if not spurious. When Italian researcher Enrico Mariutti looked at the carbon-price modeling for solar panel systems with an extremely granular focus—improvements to the efficiency of silicon ingot slicing achieved by upgrading from a steel to a diamond saw must account for the added carbon of the diamond production, for example—he concluded that the purpose of the current models seems to be "to convince the readers that the carbon intensity of photovoltaic energy is very low."[81] Mariutti unhappily figures that the carbon intensity of today's solar panels could well be ten times as high as we're accounting for today: "We are investing hundreds of billions of dollars a year in technologies that are low-carbon only because someone wrote it down somewhere."[82] How clean is a solar panel system really if it's charging a battery that was made with nickel that was excavated with power from a new private off-grid coal plant in Indonesia?[83] Behind our democratic back, capital will look for "cheap nature" to exploit, and it will find it and sell it to us wrapped in green, especially if that's the color the public is subsidizing.*

There is a further danger with the decarbonization replacement plan: What if, instead of replacing internal combustion engines, we just add electric cars on top of them? What if we do the same thing with fossil fuel energy? The simplest, most concise and consequential answer is that we will rocket past two degrees of warming with catastrophic costs. And yet so far, that has been what we've done. The International Energy Agency clocked 2023's fossil fuel emissions at 37.4 gigatons, a new all-time record high, and when you dig into why, it's clear how little humanity seems to be taking charge of its own destiny: Fluctuations in the demand for cement, the interest rate, and the varied effects of temperature changes (and the corresponding amount of heating and cooling populations require) are driving changes in emissions more than any concerted plan to decarbonize the worldwide social metabolism.[84] In fact, major producing countries are openly planning for exajoules of new fossil fuel extraction in the coming years.[85]

There's a good historical precedent for this energy substitution failure as well. In a brilliant study, Richard York analyzed the effect of the popularization of petroleum power on whale hunting. Whales were hunted for their blubber, which served as a fuel source—think *Moby-Dick*. And yet with the proliferation of substitute fossil fuels and fossil fuel technologies in the nineteenth century, whale hunting became *more efficient*. Rather than save the whales, fossil fuels gave the hunters faster boats and more deadly harpoons. The market-replacement solution chased whales to the edge of extinction before twentieth-century bans herded populations into recovery. York applies insights from the

* See Jason W. Moore, "The Rise of Cheap Nature," in *Anthropocene or Capitalocene? Nature, History, and the Crisis of Capitalism*, ed. Jason W. Moore (PM Press, 2016), 78–115.

study to the green transition, concluding, "'Green' technological fixes, such as developing renewable energy sources, are unlikely to curb environmental degradation unless there are corresponding political and social changes that ensure technologies are applied for the common good, and environmental protection is promoted as a goal in itself."[86] We've covered what some of those political changes could look like, but it's hard to escape the more general conclusion that efficiency and profit are not necessarily good proxy measures for progress toward the end the implied reader of this book hopes to achieve: a better world, for the world.

We've established that a gain in production efficiency can leave us further away from fulfilling our needs than we were to start with. This is what economist William Stanley Jevons theorized in the nineteenth century when, observing the British Industrial Revolution, he concluded that improvements in coal efficiency were leading manufacturers to *increase* their coal consumption rather than decrease it. The more efficient the engines, the better investment they made, the more people were employed, and the more coal was burned. This "Jevons paradox" makes it very difficult to translate technological progress into sufficiency. Capitalists will always face a market incentive to spend down whatever economies the collective intelligence can muster.

We can see a contemporary Jevons paradox in the "sharing economy," a thesis about the ways in which societies can reduce their resource consumption by employing internet technologies to facilitate increased utilization of their stuff. You don't need to buy a chainsaw for a single project if your neighbor has one and will let you borrow it, maybe for a few bucks, the thinking goes. The idea has inspired "buy nothing" groups, in which neighbors post their excess stuff and offer it for free; they can also post complementary requests for free items from other people. Car-sharing platforms were a big part of the early sharing economy, and they invariably made the environmentalist promise of fewer vehicles on the road. It's logical: Making carpooling and hitch-hiking easier gets more people where they're going with fewer cars. But the "need" for car trips isn't fixed, and as rideshare platforms became deregulated markets for cheap service work (subsidized by oil-capitalist investors, in fact), the demand for solo rides increased, as did the number of cars on the road.

A bunch of studies have shown that the proliferation of ride-sharing is associated with a huge increase in vehicle miles traveled, more than doubling the number in some urban transport systems.[87] One analysis of US data found rideshare platforms responsible for escalating numbers of crash fatalities: thousands of excess deaths between 2010 and 2016, thanks to the extra rides.[88] Capitalists often find it more profitable to cause new problems than to solve problems on their own terms—not a great quality in a group of people to whom we're turning for solutions to existential, planet-scale problems.

A WONDERFUL, AWFUL IDEA

Progressive marketcrafters plan to kill capitalists with kindness, to trick them into financing their own euthanasia. One day the capitalists will wake up and realize that owning the means of production isn't what it once was, and they won't be in a place to do anything about it. The class divide narrows in functional importance until being a capitalist is something like being titled nobility today—maybe correlated with positive life outcomes but probably not something you can hang your hat on, not a *living*. Capitalists, understandably, do not want this to happen, at least not most of them, and yet democracy makes the rules. If the American people turn against corporate power and elect politicians who pursue a progressive marketcraft strategy, capitalists are stuck with the stick's short end, much the way workers have been for the last fifty years. The government prints the money, writes the laws, and pays the army: If the people vote to phase out capitalist command of the economy, what are a bunch of dentists who own Exxon shares going to do about it?

When Dr. Seuss's titular Grinch decides to steal Christmas, there's a moment when his frustration with the Whoville holiday turns into a plan for action, when he decides to transform himself into the Anti-Claus. "Then he got an idea! An awful idea!" Seuss writes. "The Grinch got a wonderful, awful idea!" This is the moment when he changes from a jealous lonely Who to the sole member of a new species characterized by wrath: the Grinch. In the 1966 animated film, the transformation is physical: His eyes narrow to slits, his brows tilt down hard, and his sinister smile envelops his face—the ends of his mouth coming to rest at a higher altitude than the interior corners of his eyebrows. It is genuinely creepy in the way that only animated children's stories from the '60s can be. That is the same look that I imagine hitting the capitalist face when, pushed to the edge, that class, too, has an idea, an awful idea, a wonderful, awful idea.

The problem with capitalism is not the capitalist personnel who happen to be running the show at any given time, but that doesn't mean the capitalist personnel who happen to be running the show at this given time are not a problem. During the twentieth century's last quarter, capitalists proved they were not condemned to death by universal progress, as many people around the world had hoped. They could and did fight back as a class and take their seat atop a new international division of labor—one that, as we've seen, was predicated on the exploitation of cheap workers, cheap nature, and accelerating carbon emissions. Instead of bowing before history's long arc toward justice, capitalists rewrote the story and put themselves at the ending. If that was incompatible with democracy existing in some places, then that sucks for democracy. That's why the Grinch grins: He realizes he can break the law. But whereas he disguises himself as Santa Claus, fossil fuel capitalists dress as the whole world.

In his excellent book *Carbon Democracy,* Timothy Mitchell looks at his subject not in terms of democracy and oil but in terms of "democracy *as* oil—as a form of politics whose mechanisms on multiple levels involve the processes of producing and using carbon energy."[89] The basis for the oligarchic eighteenth- and nineteenth-century European representative governments, he writes, was the agricultural production of Europe's colonies, which provided the sun-raised calories to jump-start coal-based industrial production in the metropole. As coal workers leveraged their control over energy production for wage gains and a bigger role for labor in politics, the Grinch-grinning capitalists switched to oil. We are still living in the state system devised in the wake of World War I, a system formulated more for Euro-American access to oil than for democracy in the world. "[S]elf-determination," Mitchell writes, became "a process of recognising (and in practice, of helping to constitute) forms of local despotism through which imperial control would continue to operate."[90] Saudi Arabia, for example, is named for Ibn Saud and his family, who first signed a treaty with the British in 1915. If that sounds like ancient history, it shouldn't: As of this writing, his son Salman sits on the throne, propped up by decades of American arms sales to the regime that cumulatively reach into the hundreds of billions of dollars. Insofar as the premise of representative democracy is that a nation's people get to vote for leaders who represent their interests, oil capitalists have always found special exemptions. That's carbon democracy, and it's hard to negotiate with a player who also helps write the rules.

If capitalists didn't have any real leverage over the rest of society, if they were already functionally subordinate to democracy, there wouldn't be any need to negotiate with them about avoiding planetary disaster. We'd just tell them what to do. But by definition, they control disproportionate social resources. And though they need various government licenses and permits to operate, capitalists do *have control* over those resources in a meaningful way. That leaves them empowered to invest in more politically favorable jurisdictions and divest from politically unfavorable ones in a dreaded process called capital flight. They can even, to a certain degree, withhold their capital, making investment (and the resulting employment and economic growth) contingent on accommodations from policymakers in what's called a capital strike. We've seen the latter in the offshore wind industry, in which companies have declined to invest unless they're given more favorable terms. "We want to really substantiate the message that financial value creation is at the very core of our industry," Mads Nipper, the head of top offshore developer Ørsted, explained to the *Financial Times*. "For us that means that if we cannot get to a satisfactory value creation, we are prepared to walk away."[91] Inhuman capital can walk away from the climate crisis, but the rest of us can't, and that, too, puts us at a negotiating disadvantage.*

Putting the global green energy transition and climate-crisis adaptation plan in the hands of men like Mads Nipper—not a scientist or researcher in the Montreal Protocol mode but a former chief marketing officer for the Lego Group—is like resting a $10,000 wedding cake on a $10 folding table. But from bottom to top, marketcrafters don't seem to have another choice. Even the best-intentioned, most down-to-earth policies, such as the Whole-Home Repairs bill, have to go through owners. For renters, a new heat pump might mostly mean paying more money if and when landlords figure out the obvious way to channel the government subsidies into their own pockets. As a result of this dynamic, as well as high housing prices, we can expect many renters

* For a detailed examination of the inhibiting role profit concerns play in clean-energy investment and development, see Brett Christophers, *The Price Is Wrong: Why Capitalism Won't Save the Planet* (Verso, 2024).

to go so far as to refuse and obstruct free repairs and climate-conscious upgrades for their homes rather than risk a rent hike. The fundamental shape of American society isn't popular rule, it's class conflict, and that fact colors any policy anyone tries to make.

It would be easier if the ruling class would compete to the death among themselves. We can imagine a green bloc of firms and investors fighting the fossil fuel capitalists throughout society, and there's some indication—Michael Bloomberg's support for the Beyond Coal initiative, for one—that such a bloc could win. But capitalists are not quite as disorganized and competitive as they would have us believe. I've already mentioned corporate consolidation, the growth of monopoly power, and the ways in which antitrust visionaries are trying to fight back, yet capitalist players also collude on other levels. Sometimes that's in the form of monkey see, monkey do corporate leadership, and sometimes it's a bunch of car dealers getting together in a convention center and throwing back a round of drinks while they plan society for the rest of us.

In 2023, *Slate* reporter Alexander Sammon attended the annual National Automobile Dealers Association (NADA) convention to report on the electric vehicle transition progress. What he found—in addition to a retinue of solid B-list celebrity guests, including Deion Sanders and Nelly—was alarmingly strong hostility to EVs. That's not irrational for dealers: "Compared with traditional cars," Sammon writes, "EVs have far fewer component parts; they don't need constant servicing or oil changes. That means that electric vehicles generate 40 percent less aftermarket revenue."[92] Aftermarket revenue is where dealers clean up, and NADA isn't about to skip down the plank, even if the path is laden with government incentives. The convention's speech about the ways in which dealers can benefit from federal EV subsidies—the kind of educational marketcraft that makes the subsidies work in real life—took place in a room that was close to empty, attracting not even a handful of dealers.

NADA is a deeply political organization, because it depends on state-level laws that prevent consumers from buying cars directly from manufacturers. At the 2023 convention, former UN ambassador Nikki Haley practiced her stump speech ahead of the declaration of her presidential candidacy. She got the day's biggest round of applause when, referring to her tenure as governor of South Carolina, she told the packed crowd, "I was a union buster, and that's something I was very adamant about."[93] For car dealers, that's more important than creating a sustainable industry. At the UN, Haley helped orchestrate the American withdrawal from the Paris Agreement.[94]

NADA's practical boycott of the EV transition, subsidies be damned, is a good example of how capital acts when faced with a political challenge. "Dealers stand between many electric cars and most American car buyers, but they aren't just going to lay down and let some zero-emissions playthings roll them over. Some, I heard over and over, would rather not deal than deal with someone else's dictates," Sammon writes. "They would self-sabotage if they had to."[95]

Yet if there's a socially productive conflict between the car dealers and the battery-materials miners, it's not yet evident. That same year, the annual Prospectors & Developers Association of Canada (PDAC) conference, the world's largest such event in the mining industry, was "awash with talk of new mines and big profits," reported Nick Bowlin for *The Drift*.[96] At PDAC, everyone was happy to talk about EVs. "To stop global warming," the head of McKinsey's battery-materials team told his audience, "you need us."[97] McKinsey's planning advice to auto dealers, however, warns against electric vehicles in the near term, cautioning them with regard to the absence of oil-change revenue.[98] A 2023 report from the consulting firm suggests that most EV sales will occur in fewer than 10 percent of American "designated market areas," even in 2030, thanks in part to varied "regulatory conditions" secured by NADA-supported policymakers.[99] Within particular markets, "higher demand for new cars does not necessarily imply high EV adoption," the consultants consult.[100] We don't need to wonder whether capital can perpetuate the status quo and profit on the green transition at the same time: Individual firms already do.*

The levers of investment are powerful, but capitalists—fossil fuel, green, and the hedged combination thereof—have every reason to relentlessly grab at subsidies and thrash against restrictions each step of the way. We know from history that, in order to get a favorable outcome, capitalists will pout, cajole, influence, distract, confuse, bribe, sabotage, threaten, and even murder. If they know that the world has no other option ("To stop global warming, you need us"), they will freely invest in antisocial modes of profit in addition to if not instead of prosocial ones. Even if we can accommodate such a trade in principle, the world's creatures cannot afford such a strategy—the current prevailing strategy, it's important to note—in part because it involves *burning increasing amounts of fossil fuels.* But in a competitive environment, the idea that capitalists might cut off all antisocial avenues of investment is implausible; for capital, there's almost always somewhere you can pay less or dump more. A dollar saved via flaring or cutting wages is just as valuable as a dollar made by expanding production or reducing waste—and it's easier to get. No matter how tightly we cup our democratic hands, capital leaks through like water.

* "While Exxon says it wants to expand into new low-carbon businesses, it is not retreating from oil and gas. Rather, it plans to bolt the low-carbon business on to a growing oil and gas business." Justin Jacobs, "What Is Really Driving ExxonMobil's Clean Energy Commitments?," *Financial Times,* May 8, 2023.

Marketcraft, then, requires a supplement, a counterforce to capital that can not only exert its own leverage against the ownership class but also supplant it at the level of production if the people determine that capitalists aren't living up to their social responsibilities. It's this strategy we turn to in chapter 2—public power.

CHAPTER 2

PUBLIC POWER

I.

RACCOON LAKE

Solar and wind power are solutions, but they're obviously dependent on environmental variables, while our demand for power is comparatively constant. To bridge the gap, we need forms of power and energy storage that are immediately dispatchable, even when it's cloudy and the air is still. At the time of this writing, the United States Department of Energy counts 38.6 gigawatts of domestic utility-scale power storage.[1] Of that total, 22 gigawatts are in the form of batteries built into mountains, mostly from the 1970s.[2] What does it mean to make a battery out of a mountain? Why did they do it? And—most important—at a time when installing storage systems to hold and discharge petawatts of renewable energy every year is of existential importance, *Why did they stop?*

Elevation endows objects with potential energy, so if you climb a hill, your body mass becomes charged with power, something kids prove all the time by sledding and rolling downhill. At Cooper's Hill, in a town near the English city of Gloucester, a "cheese rolling" crowns the annual festival, but the sub-ten-pound wheel of hard cheese hurtled down the slope so fast that organizers once replaced it with a foam replica to reduce injuries. In our gravity-bound reality, elevation means energy, and the earth's undulating terrain means the world is just crackling with it. But how do we turn all that potential energy into something that's net-useful to humanity? We could try, like Sisyphus, to roll boulders up before letting them roll back down, but that's only storing muscle power, which has to be replaced with calories and requires Sisyphus to inefficiently haul his body up and down with the boulder: famously pointless. The real trick would be hitching the boulder to, say, Apollo's chariot every morning, when it raises the sun.

Boulders don't climb with the sun, but water does. The solar-powered water cycle lifts tons of liquid and pours it back down on the earth's uneven surface. In the North American West, accumulated snow deposits in the Sierra Nevada provide nearly a third of California's water supply, acting as the state's decentralized natural reservoir. To convert the kinetic energy of the water cycle as it drapes over the world into the electrical energy into which we can plug microwave ovens, we need to stick a generator somewhere it'll be forced to spin. You can try an existing waterway, but such "run-of-river" systems have limited output, which is why they're usually categorized as "microhydropower." At a large scale, the efficient answer is to cause an elevated lake of water to drain directly over the spinning wheel.

Taking real human control over such a system means being able to reverse the draining and filling process on demand. This allows you to save excess solar energy for a cloudy day by using it to refill the top reservoir. For that, you want the elevated reservoir, a lower reservoir for discharge, generators in between, and pumps to refill the top tank. With those simple elements, you can convert potential energy to electrical power and back again, cheaply, easily, reliably, and emission-free. It's called pumped-storage hydropower (PSH), and that's how you make a battery out of a mountain.

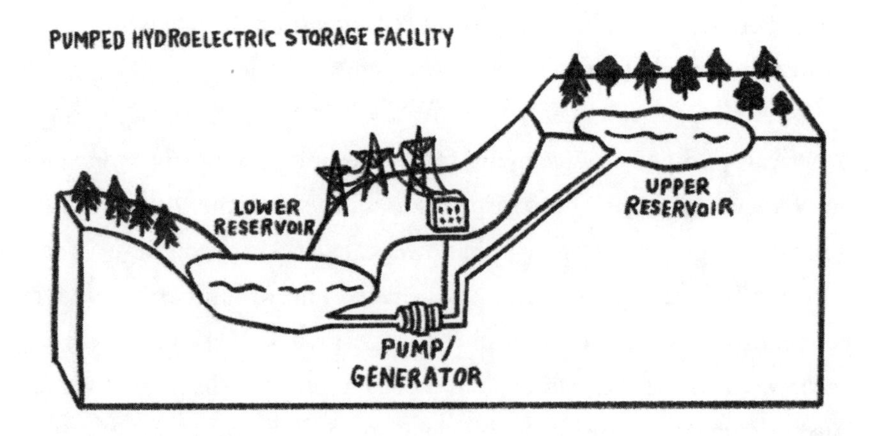

PSH provides almost all the world's utility-scale energy storage capacity. This may come as a shock even to informed readers, who are probably accustomed to hearing about energy storage in terms of lithium batteries, but the numbers aren't even close. In the United States, where installed rare-earth battery capacity nearly sextupled between 2020 and 2022, PSH still provides 70 percent of grid storage power capacity (twenty-two gigawatts).[3] And because PSH reservoirs take much longer to discharge than batteries do, pumped storage constitutes a whopping 96 percent of domestic *energy* storage capacity in practice (553 gigawatt hours).[4] The global percentage is similar, in the high 90s. Lithium batteries work for electronics and cars, but for the grid, we use mountains, and we have for a long time. And yet PSH is the green transition's red-headed stepchild, at least as far as the United States is concerned. Aside from the forty-megawatt Lake Hodges plant, built and operated by the San Diego County Water Authority, which went fully online in 2012, the United States has completed *zero* pumped-storage hydro facilities in the twenty-first century. And not only that, Oak Ridge National Laboratory's 2023 map of the PSH development pipeline lists *zero* new facilities under construction.[5] By contrast, the world leader in pumped storage, China, has 50 gigawatts of PSH operational, 89 gigawatts of new capacity under construction, and a staggering 276 gigawatts in development—compared, recall, to 22 operational gigawatts in the United States, when everything is working.[6]

This all does not seem right. In the '60s, '70s, and '80s, a new American pumped-storage hydro facility went online nearly *once a year* on average.[7] At the height of installation, in the '70s, the country added more than a gigawatt of new PSH power capacity on average every year. (Since 2003, the United States has averaged fewer than two *mega*watts per year in new projects, a number that gets worse every year there are no new plants.) If the country had merely kept that '70s pace, it would currently have an additional 62.9 gigawatts of PSH, for a total of nearly 85 gigawatts, enough to put it on top of the world. What happened?

A half hour from downtown Chattanooga, Tennessee, is Raccoon Mountain Reservoir. It's a famed recreation spot for kayaking and canoeing, fishing, hiking, and watching the local bald eagle population. It sits on a bend in the Tennessee River; tourism review sites and travel blogs are glutted with favorable reports. In 2005, the nonprofit Southern Off-Road Bicycle Association signed a land-management agreement that allowed it to build and maintain a set of mountain bike trails, where Go Nuts Biking now holds squirrel-themed USA Cycling regional championship races.* The trail names (including Megawatt and High Voltage) suggest there's something else going on at this nature preserve. With more than one hundred billion gallons in the upper reservoir, Raccoon Mountain is the country's largest energy-storage system—with more than thirty-six gigawatt hours of storage, it's second in the world, just behind the hulking Fengning pumped-storage power station in China's Hebei Province. Raccoon Mountain is a triumphant example of PSH's American golden age, and it can tell us a lot about why the technology fell off.

* The bike trails are a good example of what journalist Kate Aronoff calls "pool party progressivism." See Kate Aronoff, "The Case for Pool Party Progressivism," *New Republic*, August 15, 2023.

When Congress founded the Tennessee Valley Authority (TVA) as part of Franklin Roosevelt's New Deal, the agency was at least as much about economic development as it was about electricity per se. Left behind by private utility capital in the twentieth century's first decades, the Tennessee Valley—which also includes land in Alabama, Georgia, North Carolina, Mississippi, Kentucky, and Virginia—required direct government intervention rather than mere marketcraft. The TVA modernized the region in the original sense, moving living standards and production processes into the modern age, foremost through electrification. The TVA's first dams controlled flooding and provided power, winning over business leaders and attracting capital to the area with super-low electricity rates. In midcentury, the TVA expanded to provide power for World War II industry and afterward for a growing manufacturing sector. The TVA turned to coal plants in the '50s to satisfy growing demand, but the system was still dependent on hydroelectric dams, and its constant power generation was a bad fit for growing consumer use (think new plug-in home appliances), which tended to peak at particular times. After a small experiment at the Hiwassee Dam, in North Carolina, the TVA was ready to move forward with its first large-scale pumped-storage plant.

Construction on Raccoon Mountain began in 1970 and didn't finish until 1978, occupying more than one thousand workers who dug twelve thousand feet of tunnels inside the mountain, including a football-field-size cavern for the generator room. It's a public works project from another era, an enduring representation of what the original New Deal could do—though it was built in the '70s. In a way, we can read Raccoon Mountain as the Roosevelt agenda's last gasp; in the '80s, even the TVA went corporate, slashing jobs and focusing on the bottom line.

In the '90s, new electricity industry marketcraft separated generation and transmission in the "deregulated" regions, leaving pumped storage with nowhere to stand. Transmission organizations weren't going to pay the huge up-front construction costs on something they weren't set up to charge for anyway, and neither were private generators, who could find much cheaper paths to the same wattage. Besides, fossil fuels were already a great way to store energy. Renewable storage was not an important part of the plan, and regulatory attempts to carve out a spot for PSH in the ecosystem of green incentives have been unsuccessful. The HydroWIRES (Water Innovation for a Resilient Electricity System) Initiative from the Department of Energy's Water Power Technologies Office has put in valiant work trying to craft a place for PSH in the market since its founding, in 2019, but the fish have yet to bite. What this means is that, despite the battery hype, the United States has actually added very little grid-scale energy storage in the twenty-first century.

Pumped storage has a number of problems from the green capitalist perspective. "In liberalised electricity markets," the International Energy Agency explains, "long lead times, permitting risks and a lack of long-term revenue stability have stalled pumped-storage hydropower development, with most development occurring in vertically integrated markets, such as China."[8] As Brett Christophers writes about asset funds, you'll recall, capital looks to get in and out of infrastructure projects quickly. Hollowing out a mountain is the exact opposite of the kind of thing it wants to do, especially when some magic transformative energy storage technology is always theoretically right around the corner, threatening to disrupt all their calculations. No 2020s capitalist wants to throw down $1 billion for a ten-year bet on one-hundred-year-old technology, not based on a market framework that the authorities are making up as they go along. What if the political winds shift against the green transition halfway through construction? Or what if they shift against interfering with waterways? A small tweak to the marketcrafted incentive structure could send a project into chaos.

Even if you're looking at renewable storage, it's much better to invest in something that can scale flexibly, with lighter labor costs. Though there's still plenty of digging work involved in rare-earth batteries, you probably don't have to oversee that part, and the labor disappears into the purchase price.

From this viewpoint, it's easy to see why the United States has ceased PSH construction: Big pumped-storage hydro projects only make sense when you think about the system holistically, from a long-term perspective, the way the TVA did once upon a time. (Agency-funded researchers concluded that Raccoon Mountain would more than make up for the damage to the fish habitat with the new reservoir territory, which seems to have been the case.[9]) No matter how much we nudge and bribe and scold capitalists to go this way and not that, planning for the whole social body is not their job. That's why we need public power.

PEOPLE'S CONTROL OVER MODERN TECHNOLOGY

Public power is about more than a government plan to directly build and operate a giant network of pumped-storage hydro facilities—though I'm convinced it does also mean that. It's also more than a mere supplement to marketcraft, a "public option" to compete with private ones. Public power is a full alternative strategy, designed to resolve the contradictions at the heart of our crisis moment. Rather than break the connections in the Oil-Value-Life chain, public power proposes to loosen both sides, to deform the links so they can no longer hold the gate to the future closed. By reducing the value of fossil fuels and providing a basic standard of living for everyone, we can escape from capital's impersonal, inhuman dictates. Instead of production for production's sake, production for our use. Instead of capitalists scurrying within a maze of democratic design, the direct social appropriation of the means of production in the common interest. Publicly owned power stations, yes, but also public control of the power to put society's resources to work. The two fuse in a project like Raccoon Mountain, which uses public power to build public power and vice versa.

Some have accused public-power advocates of taking advantage of the climate crisis to push an unrelated socialistic agenda. Why would state control over the means of production make the United States more likely to meet its emissions and adaptation goals than crafting a better market for the players we already have? Should the government, with little experience or organizational capacity, really take over EV production from the automakers, for example? Wouldn't it be inefficient to hold a national referendum on what color trim options to offer on the state hatchback? There's nothing obvious to gain by shutting down the private companies and rehiring everyone at USA Cars to do the same jobs, not from a climate perspective. If our central task as a society is to build three hundred million electric cars, then it's probably best to get the autoworkers and automakers to do it. But who's to say we need to build three hundred million electric cars? What if we *need* to build three hundred gigawatts of pumped-storage hydropower? When you look at the green transition from the perspective of grid-scale energy storage, it's clear that, no matter how great the social need is, capitalists will simply decline to invest in certain kinds of projects, even with a reasonable expectation of long-term profit. And since no capitalist is ultimately responsible for society's metabolism as a whole, there's no guarantee that some decisive detail of decarbonization won't drop between them like a fly ball in front of a lackadaisical Little League left fielder.

In the final instance, it's up to the public to democratically decide what we need *and to make it happen* with whatever resources we determine are necessary. And if a capitalist owns land in what the engineers and the ichthyologists decide is the right spot for a dam and doesn't want to sell, the TVA has been willing, historically, to expropriate that capitalist. This strategy is, when it comes to these questions, similar if not the same as what some theorists have called the big green state plan, with its hallmark of "economic coordination through non-market means, that is, through economic planning," justified by the "systematic failure of market mechanisms to deliver the large-scale economic transformation that is needed for decarbonisation, and that necessarily involves shrinking rapidly and significantly some economic activities," as Daniela Gabor and Benjamin Braun write.[10] Planning decarbonization is a lot easier for a society that's not concerned with how such a campaign will enrich investors.

It's one thing to incentivize a green transition, but it's another thing to accomplish it no matter what. For public-power advocates, the electricity system is too important and close to the social heart for us to leave it to market actors. "Only publicly owned electricity can invest and plan with long-term infrastructural—and planetary—goals in mind," writes Matthew Huber in his call for public power, *Climate Change as Class War*.[11] Like Huber, policy analyst Matt Bruenig uses the TVA as a reference point for what a public-power transition would look like, and he goes as far as to suggest that the TVA itself is still the closest thing we have to an appropriate vehicle. Drawing a contrast with the marketcraft strategy, Bruenig writes, "Because it is a federally owned corporation, it can be used directly by the government to achieve energy transition goals. This differs from other approaches that rely upon subsidies and mandates to indirectly modify private sector behavior."[12] Reforms such as broadening the TVA's geographical ambit (to everywhere) and reinvesting it with congressional appropriations along with a new mandate to decarbonize would cause what he calls a green TVA to lead a public-power push.

What we as a society need or don't need is a question of values, not linear algebra or the aggregation of innumerable self-interested maximizing choices. Technology and development, likewise, are not preordained two-dimensional paths on which we can only advance slower or faster. This is not Super Mario. Public power says that we have to face choices about what kind of society we want to make head-on, not contract them out to entrepreneurs. That means thinking about what *kinds* of technologies we want to develop—not just how many dollars' worth—and what corresponding ways of living those planning choices enable and suppress. We can only begin to approach these crucial questions in this manner from the perspective of public power.

There is a perfect example of sophisticated public-power thinking along these lines in a 2023 report from the Climate and Community Project (now the Climate & Community Institute). In "Achieving Zero Emissions with More Mobility and Less Mining," the authors approach American transportation planning and EVs from a social-metabolic, public-power perspective: Instead of looking at the problem from one side, they examine a few different variables, including lithium demand, average battery size, rate of vehicle ownership, and battery recycling.* The paper offers four different planning scenarios, ranging from mere electrification to a deep transformation in land use and urban design that would increase walking, biking, and the use of public transit.[13] Profit and even cost of development are not considered. The report's real virtue isn't the revelation that the United States can deeply reduce its incipient demand for rare-earth metals by moving away from car dependence—that's simple. The point is the kind of planning-thinking that a public-power perspective enables, the kind of thinking in which a neighborhood bus schedule in California is connected to a mine in Australia. Anything less than that is willful ignorance. Among the benefits of the low-lithium scenarios, the report concludes, are mitigated harms of lithium mining,

* Obviously a state-size effort on the same question could include many, many more variables.

reduced geopolitical tensions with regard to control over mineral deposits, safer communities, and *achieved climate targets*. Such a scenario would also be in line with public power–style efforts in mining countries to capture more of the value chain—think Chile's nationalization of its lithium and Indonesia's successful ban on the export of unrefined nickel—which means richer countries paying more for less.[14] This equalization of global exchange is a basic requirement for what's called a just transition. Those are the kinds of planetary goals Huber refers to, the ones even well-intentioned capitalists are not able to achieve because they require social planning at a level that is above the capitalist pay grade.

We shouldn't confuse public power with the nationalization of the existing production system, with USA Cars. Nationalizing the US power grid, for example, could set the stage not just for its necessary expansion but also for its "rational operation" in the public interest as well.[15] By socializing the planning function, the big green state refuses to see climate goals subordinated to capricious energy executives and auto dealers. If capitalists want to come along for the ride, there will be plenty of opportunities to buy bonds from the TVA or some other green government project: Instead of the government investing in capitalists, capitalists can invest in the government. And rather than regulate noxious industries to death with the imprecise weapon of legal rule making, public power would expropriate them directly and wind down operations, evaluating previously private capacity for new socially planned use. Drilling machinery that once bored fracking holes might carve the next Raccoon Mountain, but only if we get the public's collective hands on it. The Democratic Socialists of America's Ecosocialist Working Group has called for the state to nationalize fossil fuel companies in order to phase production out and for the former executives to be tried for "crimes against humanity."[16] That is also a form of social planning.

PUBLIC LIFE

If public power were just about nationalizing the electricity system, it would be more like an addendum to the marketcraft strategy than a whole strategy of its own, the way public education and municipal waste contracting do not per se undermine the capitalist mode of production. That is one way to think of it—as the far end of what some scholars have called the "continuum of administrative control."[17] But social planning involves more variables than price per gigawatt, and public power means something qualitatively distinct from marketcraft.

Capital looks to exploit cheap nature and cheap labor, which means, as I've said, that capitalists systematically underinvest in areas of production where nature and labor are expensive, even if those are areas where we as a community of human beings are in desperate need of investment. We can see that in the manufacturing sector, where firms have chased low-paid labor and lax environmental regulations around the world, and in services, where particularly low-wage firms have triumphed by continually driving their costs down. Public power is a different kind of plan.

It's not hard to think of some areas of social life where we'd be better off without capitalists, and public-power advocates have plans to take over most of them. The private medical system is an easy one, especially after the COVID crisis dramatized just how dependent the industry is not just on public research and funding but also on genuine public planning. Looking back at Operation Warp Speed, the government program to develop a COVID-19 vaccine via private pharmaceutical companies, three researchers conclude that it was an example of how *not* to craft industrial policy because it created huge profit spikes for the companies involved without generating vaccine security for the world or even the country in the medium term. In their review, they draw a comparison with climate change, writing, "We should be skeptical of a 'silver bullet'—technological solutionism—and focus on social and political change rather than wait solely for technology to save us. But clearly there is some role for technological innovation in the fight against climate change and its effects on people and nature. And there is good reason to think that letting industry write that innovation agenda will reinforce the status quo and fail to produce the technologies we most need."[18]

In our model of cheap-chasing capital, we should expect a private medical industry to lean toward profitable medications produced in low-wage countries rather than wage-intensive preventive care, even if the latter works better and saves society money in the long run. A public-power health-care system, however, might tend to overinvest in low-cost preventive care or even in simply creating a more healthful environment, taking the onus off the individual, where it doesn't belong in the first place. Private health insurance companies are not going to invest in neighborhood basketball courts, but a smart public health system may well do just that.

Even when public subsidies do lure capital into environmentally regenerative production, capitalists always threaten to squirm away and eat their cake too by doing a lousy job. Look at tree planting, in which the prospect of carbon credits incentivized firms to finance new forests around the world, only for many of the North American ones to promptly burn down.[19] And in the global South, where many people live in and around forest ecosystems, carbon-offset forests have had negative impacts on local income and food security. Capital is already using the green transition as an excuse to take advantage of the least powerful people in the world—again. Nor can we trust capitalist accounts of their own actions. "In case after case," reported ProPublica's Lisa Song after researching subsidized forest production, "I found that carbon credits hadn't offset the amount of pollution they were supposed to, or they had brought gains that were quickly reversed or that couldn't be accurately measured to begin with."[20] In addition to the number of trees they plant, a public-power tree-planting corps might consider how to create good jobs for young people, improve landscape resilience, and make a better habitat for particular animals. To do real social planning at the metabolic level, ensuring that the things we democratically decide to make happen actually happen, we need public power ready to step in and get it done directly when we suspect capitalists will refuse or do a poor job. Otherwise we're not really making plans; we're making Amazon wish lists.

One of the main problems with the Oil-Value-Life chain is that human desperation anywhere becomes a threat to human life everywhere, because capital reliably conscripts the reserve army of labor into environmentally noxious work. To counteract this predatory (though impersonal) phenomenon, public power seeks to ensure that no one out there is left vulnerable and alone, dependent on capitalist vultures in order to meet their basic needs. Along with public health care, public housing is therefore the strategy's foundation as much as electric power is. Childcare, education, and eldercare as well. And there is no life, public or otherwise, without recreation and art. A picture of a public-powered life comes into focus, and it involves a shift of collective resources toward low-carbon, high-touch care work. With more than enough doctors and nurses to go around, capitalism can't hold your loved one's life over your head and compel you to smuggle petrol in order to get enough money to save them, for example. No one could even justify their highly paid antisocial career with their child's need for a good education, not if everyone's child was entitled to a good education regardless of parental income. Perhaps this is counterintuitive, but in such a world we could all take more personal responsibility for our choices.

With this social state layer, any number of new kinds of lives become "worth" living. In this way, we can see public power as a strategy for increasing people's genuine freedom, in contrast to a petro-capitalist vision of freedom that revolves around low gas prices, lots of roads, and an unpaid wife at home doing the dishes. That question of freedom is connected to the climate crisis because it's the lack of that freedom both at the social level (to plan) and at the individual level (to make a life without exploiting or being exploited) that has us locked in a conspiracy to destroy our own planet. There's a potentially virtuous circle, in which the more public power we can build, the more freedom we have; the less capital controls people's life choices, the more public power we can build; and so on. Public power has a revolutionary goal because it calls for a change in the mode of production, but the strategy can proceed stepwise. The quantitative accumulation of freedom eventually reaches the point of qualitative social transformation—ideally without an explosive inflection moment. We keep building, and one day we wake up and realize that we've secured enough latitude to plan and make a better world. Maybe we'll only know that things have changed for sure when we see that the level of atmospheric carbon has declined.

Public power has already begun gathering. In a bit of metastrategic overlap, marketcrafters working on the Inflation Reduction Act included a direct pay provision, allowing the federal government to pay tax-exempt entities directly rather than offering them tax credits they can't use. This opens the door for governments and tribes—including agencies such as the TVA—to compete with private generators for subsidies. It's easy to see that, with its superior financing and planning capacity, along with its public-interest mandates, public power has an advantage. (After all, the government does print the money.) From there, it's about building political will and navigating the treacherous policymaking and budget processes. It's a hard course, but early indications are that it's possible: In New York State, the DSA's Ecosocialist Working Group led a multiyear Public Power NY campaign that culminated in the successful Build Public Renewables Act (BPRA), which reversed a ban on the New York Power Authority's (NYPA) ability to build new-generation infrastructure.[21] Further, the BPRA commits the public utility to building renewables in order to meet the state's ambitious climate goals, if and when private investors fail to make it happen, a failure that is in progress at the time of this writing. If the NYPA builds public power, the state's people will get access to cheaper energy, but more important, they'll have democratic control over the energy system.

Where, though, does public power accumulate? If it were simply a matter of state ownership, we'd already be most of the way there, because the majority of the world's fossil fuels are under government control. As we saw in Chad, however, when a government takes over a national oil industry, it doesn't necessarily indicate that there's a green transition coming. Public power as a strategy involves the development of a certain kind of government, one that corresponds to a particular balance of class forces. After all, a government that doesn't represent its people can accumulate all the power in the world without elevating its public an inch. Public power has to be more than a pour of gifts from the governing heights: It's an interaction between that rain and the grass roots stretching to meet those drops in the middle.

THE OPPOSITE OF CAPITAL

If public power is the power to plan production in the public interest as decided democratically, then it involves the dispossession of capitalists. Not necessarily dispossession of any particular piece of their property but of their general prerogative to plan production for private profit. That's their social role, without which they're just idle financiers of the real leaders, as they were before capitalism predominated and moneymen had to yield to the whims of kings and queens and priests. Capitalists are willing to share a certain amount of power if it's administered by elite technocrats—technocrats they might know from school or the revolving door between government and industry—but ceding it to their enemies in the working class is another thing entirely. Unlike marketcraft, which entails a subsidy-greased class compromise, public power means class conflict. The problem with the "public" as an avatar in that conflict is that, though it technically refers to the whole polity, a small class of capitalists dominates all its representative institutions. One need only count the millionaires in Congress to understand why America's democracy does not seem inclined to support—no matter how reasonable and even logically necessary it is—a qualitative expansion of public power. If the public isn't opposed to capital, who is? Labor, when it's organized.

The cycle of public-power accumulation needs an external stimulus to get moving, and the strategists have a good idea where it should come from. Matthew Huber writes: "It takes actual working-class institutions embedded in everyday life, like unions, political parties, and the concrete processes of struggle in the workplace, to build power.... It is this kind of power—the disruptive power of workers whose own labor guarantees the profits flowing to capital—which has the capacity to 'create a crisis' for capital and force capitalists into the kind of concessions a Green New Deal represents."[22] Just as nuclear fission generates a burst of energy by splitting an atom of matter, public power splits capitalism's bonds at the atomic level, exploding the un/equal relationship between capital and labor on which the whole system is built. It only takes one split atom to start a chain reaction, and workplace explosions can happen the same way. Where to start trying to trigger a meltdown is a strategic question, and there are a few different approaches among public-power thinkers.

Huber points to the International Brotherhood of Electrical Workers (IBEW) as an already organized labor body with the ability and know-how to dispute control of the electrical grid at the points of production and distribution and, with it, the rest of capitalist society. Capitalists depend on the grid, both for their personal consumption and to literally power accumulation. For the IBEW, helping enforce a strike somewhere else in the economy might be as easy as flipping the right switch. And to the degree they're able to empower the working class to build public power, the IBEW will play an important role in planning the new system, since its members will be the ones running it on the ground—that is, once we get the lines off those damn poles and buried. Autoworkers, who are now disproportionately responsible for implementing the country's decarbonization strategy, are an important node in the network, too, and the 2023 United Auto Workers strike showed they could force their agenda on their bosses. A truly public-powerful autoworkers union would do more than negotiate the conditions and compensation for the EV transition—though that's already a leadership role: They'd also help plan what kind of vehicles to build as well as how many and why. In a world organized around public power, autoworkers in the United States would be doing that planning alongside the internationally organized mining and refining workers who produce the metals that Americans plan with. And don't forget Chinese anode makers.

For others, public power emerges organically from those low-carbon, high-touch care jobs, not necessarily from the factory floor. Writes scholar Alyssa Battistoni:

> Though the hard-hat vision of the working class retains a surprisingly tight grip on political imagination, the fastest growing sectors of the economy are in industries characterized by "pink collar" labor—nursing, teaching, service work. Green jobs boosters often note that there are more jobs in solar-panel installation than coal mining these days—but there are also

> more teachers, home health aides, and child-care providers. These jobs are done disproportionately by women, immigrants, and people of color.[23]

These workers are a natural constituency for public power, partly because they are poor investments as far as capital is concerned and partly because their crucial role in getting society from one day to the next makes them "essential." This work is sometimes described as *re*productive for that reason. Since they represent an unfavorable area of investment for capital and the state has declined to pick up the slack, job conditions and pay have steadily worsened in these sectors. When the COVID-19 pandemic hit, in 2020, some workplaces functionally collapsed because workers were stretched too thin, leading to innumerable wildcat walkouts as well as sustained union organizing efforts at Starbucks and Amazon.

Brushfire organizing across America's workplaces is hard to contain, but it can also be hard to meaningfully support. This conundrum led the Democratic Socialists of America and the United Electrical, Radio and Machine Workers of America (UE) to launch a project called the Emergency Workplace Organizing Committee (EWOC), which volunteer Gabriel Winant describes as "a kind of help desk for workers' organizing."[24] Unlike car manufacturing and electrical grid maintenance, these sectors are labor-intensive, which means organizing them organizes a lot of people, which is its own kind of class leverage. In February of 2020, right before the pandemic hit, thousands of unionized janitors in Minneapolis went on strike with workplace and climate demands, including the shutdown of a local incinerator.[25] Teachers unions have adopted a similar practice, making broad social demands in addition to workplace demands—sometimes called "bargaining for the common good."[26] By linking society's refusal to invest in their work with capital's insistence on investing in fossil fuels, organized care and service workers demand a planning role for labor, which is another way to say public power. Every time workers win, they are exerting control over re/production, and with enough wins, that control assumes a qualitatively new form. As Battistoni writes, "Rather than letting the private sector alone determine which kinds of work are worth doing, a program for environmental remediation must assert a strong public role in creating jobs that improve human life and environmental health, regardless of whether they also generate profits."[27] Not "demand," mind you, but "assert."

A third, overlapping view holds that working-class institutions are compromised by their association with the Democratic Party, which is fundamentally a capitalist class-conciliationist tool. At best the Democrats might be cowed into some useful marketcraft, but many unions have committed themselves to a political agent that works, overall, against their interests. Organized workers need to stop wasting their time, break up with the Democrats, and form an independent party that will actually accumulate public power rather than compromise it away. Labor Notes cofounder Kim Moody explains this tendency well in his book *Breaking the Impasse,* where he describes labor leadership's hesitancy to "kill the (capitalist) goose that lays the golden egg."[28] By splitting with the Democrats, labor is free to focus on expanding its ranks and organizing the unorganized rather than on electoral campaigns. "Working-class power does not derive from holding office in the capitalist state even if that can enhance its ability to win things under some circumstances," he writes. "The point is to activate that power at its sources."[29] If the working class ever has the power to front the public and its political institutions, it won't be behind the Democrats; it'll be in labor's own name.

What these examples show is that organized labor is both the subject and object of the public-power strategy. The working class builds public power, and public power builds the working class. Each expresses itself in the other. If that sounds complicated conceptually, it's not so complicated in practice. An unleashed TVA with a mandate to build and operate three hundred gigawatts of pumped storage means planning to create a lot of new union jobs. Those workers add to labor's organized capacity, which puts the campaigns to replace a bunch of fossil fuel plants with public renewables over the edge; that in turn means a lot of new union jobs. This dynamic can spin at the smallest public level: School districts across the country have installed solar arrays, saving money but also *building power* that they've used, for example, to increase teacher salaries, provide green job apprenticeships, and distribute free Wi-Fi access to the community.[30] Economist Lenore Palladino suggests that unions of public workers speed the process by using the power in their pension funds to finance public solar arrays. New York City's public worker retirement funds already hold nearly $6 billion in infrastructure investments; why shouldn't workers invest in their own power?[31]* Capitalists do. We do this work a disservice if we don't recognize it as social planning.

What kind of world, then, is public-power planning?

THE BUS AND THE STORM

In order to employ marketcraft, it's not necessary to change people's experience of the world in the near term. Eventually we get to Alice's Farm, but the marketcraft strategy is more concerned with making sure people feel like things are not changing too much or too fast. Public power, however, imagines subtle but important differences in everyday life. Not too much or too fast, either, but the line is in a different place. While marketcraft proudly finds its form in policy briefs, it's wrong

* For more on the history of public-public "fiscal mutualism," see Michael Glass and Sean Vanatta, "The Frail Bonds of Liberalism: Pensions, Schools, and the Unraveling of Fiscal Mutualism in Postwar New York," *Capitalism: A Journal of History and Economics*, May 25, 2020.

to treat public power the same way. There is a romantic element to the strategy, and we shouldn't neglect it. To that end, I'll analyze a film and a novel—the first contemporary, the second from the New Deal era.

The 2016 film *Paterson,* written and directed by Jim Jarmusch and starring Adam Driver as the quiet titular everyman, doesn't have any of the so-called solarpunk aesthetics common to post–Green New Deal future imaginaries. There are no giant solar arrays or urban forests. The set is, in fact, merely the town of Paterson, New Jersey. At first glance, nothing about the material world suggests that the story takes place in a better future, but I'm sure it does. There is a bus: Driver's Paterson drives it for Paterson, making him not just a public worker but also a green worker in the field of transportation. If personal EVs are marketcraft's signature motorized vehicle, then public power's signature vehicle is the bus, in the spirit of the Climate and Community lithium report and all the other reasons to plan for fewer cars. Driving the bus fits Paterson's spirit, and he takes in parts of the city and its people as he helps knit the two into a whole from behind the wheel. He's a poet who drives a bus more than a bus driver who writes poetry, and his job gives him plenty of time to pursue his passion.

Well, not so much pursue as just *do* it: His wife, Laura, his biggest fan, can't persuade him to publish or read his work, but that's okay. Paterson's not an antisocial guy, and his routine includes a walk with his dog to the neighborhood bar, where he has a beer and passes time with the other regulars. Laura is more of a free spirit—making art out of everything, taking up the guitar on a whim, baking tray after tray of cupcakes for the local market.* Where Betty Friedan might have seen a housewife dilettante, Golshifteh Farahani plays the role differently, as someone who is loved and loving and free. There are problems

* Her practice of painting on everything in the house, including the shower curtain, reads less to me like the work of a frivolous manic pixie dream girl than a reference to the Filipina painter Pacita Abad, who similarly made art out of her domestic objects without bothering to remove them from use.

in *Paterson:* Rizwan Manji as Paterson's bus colleague Donny makes a hobby of complaining about hassles and ailments; William Jackson Harper as Paterson's bar friend Everett is lovesick; the bus breaks down, and Paterson's notebook gets eaten by his dog. The world is hardly perfect, but there's a calm to life. Characters work less than we're used to, they buy less, and they drive cars less. There is no sign that racial division exists. And yet we do not see Paterson and Laura's life as "less than" because they only have one car and cook their own food and entertain each other with their art instead of by watching Netflix on a big-screen TV. Compared to today's American average, their lifestyle looks positively austere when abstracted to its resource consumption, but their lives are staggeringly full in all the qualitative ways we really care about. Everywhere in *Paterson,* people are free to channel the world's poetry in their own particular manner.

Paterson depicts something like the variety of lives worth living under public power and how they fit loosely together in a free society. It's a deeply political film, even though (and perhaps because) the conflicts are existential. To live in struggle with where to put a line break rather than with the exploiter class over control of production: That's the goal. Public power is the bus we ride to get there together. That's my idea of a realistic public-power idyll, and it's not such a high bar.

Or is it? *Paterson*'s full tranquility comes from the syncopated rhythm of Paterson's quiet, steady routine and Laura's eclectic whimsy. Their two-piece domestic band finds support in an invisible baseline of public power that has planned for the possibilities of their lives in general, though not the beautiful shape they make of their own in particular. When the environment is predictable—the safety net of a useful, secure job for the public at a living wage, say—people have the freedom to riff. But where in the real world should we expect to find a predictable environment? Climate change is already the reality, and that change is anything but predictable. We are beset with emergencies across the socioecological spectrum, from fires to compelled migration to floods. Where marketcraft focuses on the technocratic level of disaster response, public power attends to the everyday workers who are responsible for reproducing society, even and especially when it's in peril. We don't see Paterson forced to become a hero, forced to plan and drive an emergency evacuation route during a flood, for example, but in a public-power regime, that's an important part of his role.

In the 1941 novel *Storm,* by George R. Stewart, an emergency brings the public to the fore. Set during the New Deal era, *Storm* is the account of a storm named Maria as she hits California and therefore the public of California. Stewart is heavy on scientific detail, and fittingly, *Storm*'s human story starts in the Weather Bureau, in the analysis of a junior meteorologist. Characters are as often known by their jobs as their names: J.M., or L.D., for the utility's power load dispatcher. As the storm hits, Stewart follows it through the public-power network as the workers themselves plan the real response. In doing so, he frequently contrasts these rank-and-file workers with owners. Consider the scene at the Power-Light Building when the others see the L.D. arrive:

> In their lives the President of Power-Light was only a vague "big-shot," but the L.D. spoke with the voice of God. In this opinion they did not differ from some thousands of other employees in the far-reaching system—linemen, operators, ditch-tenders, switchboardmen, foremen, even superintendents of substations and power-houses. ... During the fire at North Fork Power-House a Mexican laborer ... had spontaneously called for help to the L.D., instead of the Virgin.[32]

No matter what the marketcrafters write in their briefs and budget requests, it's those workers who ultimately wield the power. Stewart's

L.D. is a real social planner in the final instance, dispatching the power to the place where it's needed. "Through his mind were rushing figures which represented time and space modified by conditions of topography, season of year, and efficiency of men. From long skill the calculation was so rapid that it resembled intuition..."[33] What appears as the godlike power of a single man is in fact the reflected insights of those thousands of others, beamed toward the L.D. the way the sun's rays gather on a collector in a concentrated solar facility. He's not the boss or the owner; he's the L.D. And as the storm starts taking down electric lines and other infrastructure, it's the L.D. and other public workers like him to whom society looks to plan and coordinate and maintain that freedom-generating baseline of power.

Not all that planning assumes the L.D.'s Olympic aura, and people often encounter even the best of it as a hassle in *Storm*. When Mr. Reynoldhurst ("a very great man in the domain of petroleum") has his phone call drop off for twenty-five seconds, he's perturbed. He can't imagine the pole that went down or the repair crew sent to fix it ("A few minutes later, merely by using their instruments, without even stepping outside, they had located the break within a quarter of a mile") or the telephone operator as she enforces the first come, first served rule and deftly reroutes calls around the break without regard for the threats of "big small-town businessmen" or "rich women"—all the "egoists who thought their own business was always most important."[34] Public power is the imposition of planned fairness on those people, and Stewart skillfully describes it at the level of social metabolism: "As each circuit was re-established, the testboardmen in Sacramento discovered the hum of life, and then San Francisco put the circuit back into use where it was most needed. The situation became easier."[35] The situation became easier. What else could we ask for at this moment?

When we think of bureaucracy, perhaps we think of Franz Kafka's *The Trial* or an unpleasant experience at the Department of Motor

Vehicles. But to embrace the ethos of public power is to recognize what the writer Italo Calvino once described as "the amorous passion that, though unconfessed, makes clerks' hearts warm, once they come to know the secret sweetness and the furious fanaticism that can charge the most habitual bureaucratic routine, the answering of indifferent correspondence, the precise keeping of a ledger."[36] Not everyone has to obsess about air-pressure fronts like the J.M. or electrical lines like the L.D., not as long as one or two of us are and not as long as they're looking out for and with the rest. The responsibility of each is the freedom of all; the freedom of each is the responsibility of all. That is public power.

Now, what's wrong with it?

II.

THE PROBLEM OF THE INDIAN

One of the most promising new American pumped-storage hydro facilities is supposed to be located on the west side of the Hawaiian island of Kauaʻi. The West Kauaʻi Energy Project was first spun up in 2012 by the Kauaʻi Island Utility Cooperative (KIUC), the community-owned utility, which took over the island's distribution functions in 2002. KIUC's co-op structure incentivizes it to take a holistic planning view and to prioritize low customer bills, which in turn has incentivized it to look to the sunny, hilly island's natural features. The utility has endeavored to reduce Kauaʻi's dependence on imported fossil fuels by commissioning large solar plants, including the first of Tesla's utility-scale solar-plus-storage battery setups.[37] This will allow Kauaʻi to go from getting almost all its energy from fossil fuels to the reverse; with the West Kauaʻi Energy Project, the island projects it will operate on 90 percent renewable energy. On sunny days, KIUC's solar panels and batteries already cover 100 percent of energy usage, and there are a lot of sunny days on Kauaʻi. The trick is storing enough energy for peak hours when the sun isn't shining, for which we know pumped storage is an exceptionally good solution. As far as the renewable-energy community is concerned, Kaua'i and the KIUC are the American gold standard.

A co-op utility does not itself create a cooperative society, but KIUC's proactivity suggests a near-term path for building public power. And yet at the time of this writing, more than a decade since the first proposal, construction hasn't started on West Kaua'i. This isn't a case of capital flight; rather, environmental groups backed a challenge to the plan, obstructing its progress. Why? Unlike many of the most objectionable green-energy facilities, West Kaua'i has been planned with the social metabolism in mind. The facility would have recycled plantation-era water ditches, which divert water from the Waimea River. On discharge, the water would irrigate 250 new homestead farms and pastures, which would also be entitled to energy and road upgrades on the KIUC's dime. It's vitally important not to waste the fresh water, which is in low supply, and the KIUC has committed to making sure every diverted gallon gets put to use. This all sounds like the kind of ecologically informed public-power planning environmentalists have asked for, similar to the way the KIUC uses hundreds of sheep to keep solar panels clear of vegetation at the giant solar farm in Lāwa'i, built on a former sugar plantation.[38] But there are dangers to building a public-power system in the footprint of colonialism.

"I thought it was time to let the river heal," Kawai Warren told the *Honolulu Civil Beat,* "but now they want to continue doing what the plantation did for 100 years."[39] Warren is a longtime subsistence fisherman and a leader of Nā Kia'i Kai (sea protectors), one of the groups blocking the West Kaua'i project. Their fear is that the PSH plant would further entrench poor environmental planning from the previous century. Advocates were skeptical of the KIUC's plan to reuse water, pointing to a lack of farms ready to receive the millions of gallons. Existing taro farmers who rely on the Waimea's natural flow worried that the plan would be at their expense and the expense of traditional farming practices. And if the West Kauaʻi water did find its way to the sea, it threatened to drag legacy agricultural chemicals from the plantation days with it into the shore ecosystem. Pōʻai Wai Ola, the Hawaiian name for the West Kauaʻi Watershed Alliance, which organized the campaign, asserts members' claim to the land in terms that public power might not recognize: "They rely on, use, or seek to use the water resources affected by the proposed project for a host of public trust uses including, but not limited to, fishing and gathering, kalo [taro] farming, recreation, research and education, aesthetic enjoyment, spiritual practices, and the exercise of Native Hawaiian cultural rights and values."[40] That last word is important: values. How can public power, with its *public* values, reconcile with particular Native values that are not shared with the general public?*

For some public-power advocates, there's an easy answer: It can't. Public planners shouldn't privilege farmers on a particular side of an island just because that's where the river used to go. Native or Indigenous claims to the way things used to be don't measure up against our collective need to organize our social metabolism along sustainable lines. Maybe the two overlap, but when they don't, the public interest controls. There is a reflexive antilocalist strain to public power that considers commitments to individual pieces of land and bodies of water as obstructions to rational planning. They're tempted to view these Native claims as parochial, analogous if not actually similar to, say, European anti-immigrant sentiment. But this would be a serious mistake. To

* In December of 2023, environmental advocates and KIUC came to an agreement to scale down the project and forgo the water diversion component. But West Kauaʻi is not an exception: In February of 2024, the FERC denied permits for the Black Mesa Pumped Storage Project after objections from the Navajo Nation in Arizona, announcing a new policy of not approving permits for tribal lands without tribal support. "Federal Officials Reject Three Huge Arizona Pump Storage Projects Targeting Black Mesa," Center for Biological Diversity, February 15, 2024.

understand why, we need at least a functional sense of what Indigeneity actually means.

One of the most useful and enduring definitions of Indigeneity comes from a group that public-power advocates should respect: the International Labour Organization (ILO). The ILO's Indigenous and Tribal Peoples Convention of 1989 offered this two-part definition:

> (a) Tribal peoples in independent countries whose social, cultural and economic conditions distinguish them from other sections of the national community, and whose status is regulated wholly or partially by their own customs or traditions or by special laws or regulations;
>
> (b) peoples in independent countries who are regarded as indigenous on account of their descent from the populations which inhabited the country, or a geographical region to which the country belongs, at the time of conquest or colonisation or the establishment of present state boundaries and who, irrespective of their legal status, retain some or all of their own social, economic, cultural and political institutions.[41]

What this definition makes clear is that the term refers to *currently existing communities of people,* not some abstract property rights located in an individual's identity. At this point in the book, I'm hopeful that the reader will see the series "social, economic, cultural, and political" and recognize that we're talking about social metabolism. Implicit in those distinct—though not independent—Native social-metabolic orders are distinct sets of values. These Indigenous values necessarily clash with capitalist Value and the world it silently compels to exist, but they often clash with public power as well. The Tennessee Valley Authority dug up thousands of Native burial sites during its construction projects, for example, and distributed bones and artifacts to regional settler institutions such as museums and universities. More recently, it took the TVA nearly fourteen years just to count the 4,871 people whose remains the federal agency has claimed to own as property.[42] Finally, in 2023, the TVA invited affected tribes to apply to recover what's left of their relatives.

It would be a mistake to treat the conflicts between the public-power agenda and Indigenous peoples around the world as an ethical failure of public-power advocates to be rectified with consciousness raising and recognition. "The tendency to consider the Indigenous problem as a moral problem embodies a liberal, humanitarian, nineteenth-century Enlightenment conception," complained the radical Peruvian writer José Carlos Mariátegui in his essay "The Problem of the Indian," first published in 1928.[43] Rather, he writes, the titular problem is about land. Attempts to transcend this contradiction with a "plurinational" orientation that acknowledges Indigenous nations within larger national polities haven't been able to overcome the material land-based conflicts between those nations. As public-power advocates are coming to discover, not only has the "problem of the Indian" never been resolved, in the midst of a twenty-first-century global climate emergency, it also threatens to erupt into crisis.

Conflicts between Indigenous communities and state development planning—and energy projects in particular—are nothing new. On the contrary, they're a consistent feature of *modern* politics. Shell's oil extraction in Nigeria pushed the Ogoni people into struggle, leading to the execution of nine Ogoni men in 1995. A proposed hydroelectric dam in northern Norway forced the Sámi people to confront the left-wing government's public-power plan in the late '70s and early '80s. I could spend the rest of the book listing examples of these fights from the twentieth century, but I think it would be more useful to stay even more contemporary. With ecological environments around the world thrust into emergency, we should expect communities of people who are committed to living in and protecting particular pieces of land to find themselves in the same dire situation. This has led to breaks between these communities and left-wing labor-aligned settler political parties. For example, in the early 2000s in Aotearoa (New Zealand), a conflict over Indigenous versus state control of the island's foreshore and seabed led Tariana Turia to resign her position in the Labour Party and help found the independent Pāti Māori. In the latest parliamentary election, Te Pāti Māori won four seats traditionally held by Labour.[44] American readers might be more familiar with the 2016 Dakota Access Pipeline protests led by the Oceti Sakowin, which pitted tribes and their allies against the Barack Obama administration.

One way we can know that the "problem of the Indian" is a current rather than an ancient social issue is that, like capitalism, it's global. In her book *Resource Radicals: From Petro-Nationalism to Post-Extractivism in Ecuador,* Thea Riofrancos details the emergence of a fracture between the Confederation of Indigenous Nationalities of Ecuador and the government of Lucio Gutiérrez that culminated in the 2006 presidential candidacy of Kichwa politician Luis Macas under the banner of the independent Pachakutik Movement.* "In this critical juncture," she writes, "key components of what had become a militant opposition to oil extraction among Indigenous communities in the southern Amazon gained salience: extraction as ecologically and culturally destructive; *el territorio* as socio-natural space and the site of Indigenous sovereignty; and local communities as empowered under constitutionally recognized plurinationalism to veto extractive projects."[45] Nationalizing resources has often meant trying to pump and dig public power out of Indigenous lands, and Native peoples around

* For more on the context, see Raúl Zibechi, "The Art of Governing the Movements," chap. 16 in *Territories in Resistance: A Cartography of Latin American Social Movements* (AK Press, 2012), 268–98.

the world have responded with their own forms of counterplanning. One scholar argues that environmental effects from the twenty-first-century shift of China's polluting industry into the Xinjiang Uyghur Autonomous Region—the global green boom's dirty coal heart—has been an underdiscussed cause of Indigenous Uyghur uprisings.[46]

Mining in general is an area of collision between public power and Native land protectors. No matter which strategy we get behind, and even if we pick buses over cars, there's an urgent global need for large amounts of particular metals. For countries with sizable deposits, this presents an opportunity to build public power by nationalizing the resources and planning extraction in the people's interest, with derived revenues going to good jobs, social welfare, and green development. But that plan involves sabotaging whatever Indigenous social metabolism already exists in the extraction area, sacrificing a particular zone in the public's name. As Martín Arboleda writes in his book *Planetary Mine: Territories of Extraction Under Late Capitalism,* intensified extraction in Latin America has led to new-old forms of contestation, as "indigenous peoples, peasants, women's movements, and migrants, among other 'motley' fragments of the subaltern and laboring classes," resurrect communal forms of resistance.[47] He points in particular to the 2006 occupation of the Escondida copper mine in Chile.[48] It's hard to plan for an energy transition without copper.

At the same time, Native resistance is not and never has been conservative in the sense of seeking a return to the past. As Mariátegui wrote in another essay, "The revolutionary *indigenistas,* instead of a platonic love for the Inca past, manifest an active and concrete solidarity with today's Indian. This *Inidigenismo* does not indulge in fantasies of utopian restorations. It sees the past as a foundation, not a program."[49] So it is with the movements of today. That lends them a somewhat orthogonal relationship to settler campaigns to *conserve* lands, too. Perhaps the most famous example is global settler-state hostility to

Native forest maintenance practices that involve controlled burns, but there are many others. Animal conservation sounds like good ecological planning, but in practice it often means bringing the weight of the authorities down on Native hunting and fishing practices. The Sámi fisherman and political leader Áslat Holmberg has described neocolonial Scandinavian fishing laws that "make us tourists in our homelands and waters" in the name of conservation.[50] To prevent poaching, the seemingly innocuous World Wildlife Fund has conducted what amounts to a paramilitary war on Indigenous peoples who live near state-claimed "national parks" in India, Cameroon, and Nepal.[51]

Even new green-energy generation has led to Native-state conflicts, pitting Indigenous peoples against the green transition itself. That describes the fight over the West Kaua'i project, but there are plenty more examples. The Lenca people in Honduras defeated the Agua Zarca dam, a series of hydroelectric plants that threatened the flow of the Gualcarque River. After the highly publicized assassination of Lenca leader Berta Cáceres, it was capital that dried up instead.[52] But this wasn't a case of private interest run amok: The main financiers for Agua Zarca were the Dutch state-owned bank FMO and the Central American Bank for Economic Integration—public-power institutions. Sámi reindeer herders are locked in a struggle with the Norwegian government over Fosen Vind, a wind farm co-owned by Statkraft, which is of course wholly owned by the Norwegian *stat.* This might sound like not-in-my-backyard (NIMBY) antidevelopment thinking, but the Sámi are not trying to protect low-density property value; they're trying to protect their *values,* which are grounded in particular pieces of land. "If you just look at a map or aerial photos, the northern regions are relatively sparsely populated and there is not that much infrastructure," Sámi leader Holmberg told the press about the *statsperspektiv,* rejecting the NIMBY label. "To their eyes, it looks like there is a lot of empty space and could easily fit 150 wind farms."[53] But it can't, not without wreaking havoc on the complex natural and social-metabolic systems that constitute Sápmi as a living place. Despite the reindeer herders winning a ruling from Norway's Supreme Court, Fosen Vind remains in operation—both against the law and owned by the state.

If public-power planning isn't able to incorporate many, many Indigenous value systems, and if its advocates and practitioners adopt a SimCity colonial view in which land and water are fungible resources to be shuffled and sacrificed according to shallow calculations of the public interest, the strategy will bog down at best. At worst, it will end up destroying the world in all its particularities in the name of saving the world in general.

SUPEREXPLOITATION

How would the world's people vote on a public-power green transition? If we were just asking for a show of hands with regard to who wants collective control over social planning by and for the people and our continued pleasure-seeking life on this planet, and if we assume all the people vote in their own interest, it's hard to see how the strategy doesn't pass in principle. The global working class is much, much bigger than the capitalist class, so this would more or less be a referendum on our own power. The scenario boils down to something like a tautology: "Do you want more or less?" Well, more, obviously.

However, we do not currently vote as a species—those of us who are allowed the privilege in the first place. If we want a *realistic* appraisal of world democratic intention regarding public power and how it relates to the balance of social forces, we'd have to weigh people's votes according to their place in the world hierarchy. For the sake of mathematical simplicity, let's give everyone in the world a fraction of a vote based on their position in an imaginary power-based ordering of every individual on earth. The most powerful person in the world gets a whole vote. The second gets a half vote, the third a third, and so on. Hardly a perfect system, but closer to the reality than the fantastic hippie universalism of one world, one person, one vote. In such an election, public power would never stand a chance: The *four* most powerful voters alone hold a permanent majority. In democracy, inequality still goes a long way.

The world is a jumble of these kinds of power distributions—some fully equal (humanity), some very unequal (wealth), and others in between. Every person does have an inalienable ability to participate in politics and influence the greater social metabolism simply by virtue of being born with will. No human is a toaster, and to exercise power over anyone in any way invites the possibility of disobedience and struggle. At the same time, people's varied abilities to affect the course of society in their own interest is—on a statistical level, if not on the level of individuals, who can occasionally defy the odds—colored by their (frequently contradictory) structural positions. Maybe that sounds too abstract. Here's a concrete way to put the same issue: How would an *American* worker vote on a global public-power green transition?

For advocates, you'll recall, organized workers are the combined subject-object of public power: They act to transform their own collective place in the world. Workers have nothing to lose and everything to gain by pushing out the capitalists and taking hold of society themselves, in their own shared interest. But that's not quite true for *American* workers, not as such. There are privileges—possessed by many, and theoretically available for each, if not all at the same time—to being an American, even for a worker, and the rise of the global working class does endanger some of those. Though public power claims to offer every worker an increase in authority, it's not clear how that squares with the disappearance of international hierarchy or, for that matter, the disappearance of patriarchy and racism and other forms of exploitation and inequality that advantage some and disadvantage others by cutting across traditional class lines. Members of the American home-owning white male working class are exploited every hour they spend laboring, but that doesn't necessarily mean they're interested (in the strict economic sense) in the large-scale revaluation of social values that public power entails. Maybe they do not like taking the bus and prefer driving their cars. Maybe they like living in a single-family house in a neighborhood of single-family houses and do not want to be planned into a large eco-friendly apartment building under any circumstances. They can't help but appreciate the low prices Americans pay for food, courtesy of an unequal global division of labor and extractive farming techniques.* To assume that because workers as workers have an interest in their own power they will support that society's transformation is to overlook all the other ways we plug into the social metabolism.

Whether we're looking at international relations or household relations, capitalist Value structures our metabolism. The relationship between, say, a husband and wife, or an American worker and a Brazilian worker, has more to do with Value production than we might believe if we viewed them through the simplifying dyad of capital versus labor. Value isn't a sinister octopus reaching its tentacles of compulsion into unrelated areas of society; Value is the water we're all swimming in. Flopping up onto the mudbanks of public power would be a major evolutionary step for humanity, but it's silly to pretend there's no reason for working people *not* to do it. Capitalism's existing terms are still the terms under which we each make our existence, however imperiled.

Capitalists currently plan global production in part for the unequal enjoyment of Western consumers. Relative to workers in the rest of the

* A recent study of the electoral impacts of a Dutch policy that increased household natural gas taxes to subsidize renewables found that "after the policy change renters with individualized utility bills became 5–6 percentage points more likely to vote for the radical right compared to renters with utilities included in their rents." Erik Voeten, "The Energy Transition and Support for the Radical Right: Evidence from the Netherlands," *Comparative Political Studies,* March 11, 2024.

world, workers in the North Atlantic work less for more. That's not a moral or even controversial statement; it's simply the reality of international wage discrepancies. Though it's not a statistic we read about in the mainstream press, the World Bank keeps track of unequal exchange with a metric called "labor's terms of trade" (LToT). To measure that relationship, analysts compare how much labor is embodied in $1 million worth of a country's imports with how much labor is embodied in $1 million worth of that country's exports. A ratio of 1 means a country is exchanging its labor equally with other countries; a ratio greater than 1 means a country is higher in the international pecking order and vice versa. Scholars often talk about this regional composition in terms of capitalism's "core" or "center," where wages and consumption are relatively high, and the "periphery," where wages and consumption are relatively low.* As a global winner, the United States maintains an average LToT considerably larger than 1, achieving its peak in the early 2000s at above 5.[54] While economic growth isn't necessarily a zero-sum contest between countries, labor terms of trade is. No nation demonstrates that better than the People's Republic of China.

In his book *China and the 21st Century Crisis,* economist Minqi Li looks at the nation's transformation from 1990 to 2012 based on its labor terms of trade with various global regions. In 1990, the country had an average LToT of less than 1 with every region.[55] But over the course of the '90s and '00s, the PRC pushed its LToT steadily up, especially in Asia and sub-Saharan Africa. By the early 2010s, China's LToT with South Asia was over 4, and by mid-decade the country hit parity with the Middle East–North Africa and Eastern Europe–Central Asia, thanks to both smart marketcraft and the repression of working-class consumption (which allowed for the increased reinvestment of surplus in expanded production).[56] As I've pointed out, much of that growth

* See, for example, Egyptian economist Samir Amin's formulation in *Accumulation on a World Scale* (Monthly Review Press, 1974), 37–38.

was driven by an ecologically unsustainable coal boom, and as China's workers demand their fair share—within both the international division of labor and the national division of income—it's not clear where it's going to come from. Li writes:

> Within the capitalist world system, there is not another large geographic area that can substitute [for] China and generate economic surplus on a similar magnitude. The remaining large peripheral areas in South Asia and Sub-Saharan Africa have been under intense exploitation by the capitalist world system for about two centuries. Widespread political instability affects both regions. Ecologically and socially, the two regions may have reached the limits in bearing the rising burden of global capitalist exploitation.[57]

Even without reparations for past inequality, a public-power shift that rapidly equalized the international value of labor would mean rich countries voluntarily reducing their LToT and surrendering their comparatively high living standards. Internationalist public-power advocates suggest there will be plenty of surplus to go around, but if workers in the core act like they have something to protect in the current system, it's probably because they do. And if workers in the semiperiphery act like they might have something to gain through international competition, they might be right, too. The question Li asks is: If China has something to gain, at whose expense will it be? The United States has tried to make sure the answer isn't the United States by kicking down at China on the ladder of technological development and subsidizing domestic production. But the CHIPS and Science Act, for example, has quickly run into labor problems: The expansion of the Taiwanese firm TSMC into Phoenix, Arizona, has had trouble getting relatively coddled American workers to perform with sufficient vigor.[58]

Meanwhile, other national capitalists have been happy to embrace the opportunity of Chinese elevation. In Brazil, Chinese demand for beef, soybeans, and iron ore has driven a "reprimarization" of the economy, leading to backtracking on both the economic and environmental Kuznets curves (see page 21), especially deforestation in the Amazon as well as an empowerment of the reactionary ruling-class fragment of robber-extractionists.[59] Will Chinese workers fill the streets to demand that their government stop trying to increase the national share of global production in the name of protecting a foreign rainforest? Will Americans? The public-power strategy requires it.*

An incisive reader might already know what's next: We have unequal labor terms of trade *within* national communities as well as between them. For as long as industrial capitalism has existed, working classes have been internally divided. Political scientist Cedric Robinson described this racial aspect of capitalism among the nineteenth-century English working class: "The English working class was never the singular social and historical entity suggested by the phrase. An even closer study of its elements—for we have merely reviewed the more extreme case with the Irish—would reveal other social divisions, some ethnic (Welsh, Scottish, and more recently West Indian and Asian), some regional, and others essentially industrial and occupational."[60] Rather than erase differences between workers, Robinson writes, "the dialectic of proletarianization disciplined the working classes to the importance

* I strive to keep my critique mostly out of the psychological realm, but here it's worth citing Anselm Kizza-Besigye, who argues that Western horror, even on the political left, at the idea that our access to particular consumer items might decline in a better-planned world is a reflection of geopolitical castration anxiety. Taking up the Western addiction to what he (as a Ugandan) sees as our "sad bananas," he writes, "They can envision no good substitute for their sad bananas because the desire Westerners feel for them is not for the fruits themselves—even at their best bananas are a mid-tier tropical fruit—but for the relations of unequal exchange that enable Americans to take a disproportionate share of the world's agricultural surplus and, with perverse glee, waste almost half of it." I wouldn't know how to test such a claim, but I find it quite believable. Anselm Kizza-Besigye, "Banana Republics," *Africa Is a Country*, September 11, 2023.

of distinctions: between ethnics and nationalities; between skilled and unskilled workers; and...between races. The persistence and creation of such oppositions within the working classes were a critical aspect of the triumph of capitalism in the nineteenth century."[61] Instead of having to double down against a united working class, the bosses have shown themselves able to split their opponents.

Maintaining differentials between groups of workers within a polity helps keep labor costs down in general, and it gives one fraction of the class a bonus—and an interest in protecting that bonus. "In the same way that some individual workers gain advancement on the job by currying favor with the employer, white workers as a group have won a favored position for themselves by siding with the employing class against the non-white people," explains historian Noel Ignatiev. "This favored status takes various forms, including the monopoly of skilled jobs and higher education, better housing at lower cost than that available to nonwhites, less police harassment, a cushion against the most severe effects of unemployment, better health conditions, and certain social advantages."[62] We don't need to resort to psychologism in order to explain the white worker's interest in whiteness. A public-power program that agrees to maintain these racial and international inequalities among workers at an attenuated level dooms itself by half measures and to an unreliable base. Though it might seem easier to try to avoid the question altogether, rallying workers to surrender their "wages of whiteness" in the name of planned equality has to be a public-power priority. Advocates should be very careful not to accidentally reinvent national socialism.

There is an important division among workers that is notably absent from Cedric Robinson's list: gender divides in the working class, even within a single household. In her book *Patriarchy and Accumulation on a World Scale,* Maria Mies describes a "*de facto* class division between working-class men and women."[63] Like Robinson, she draws her narrative back to the emergence of the industrial working class in Western Europe. "The process of proletarianization of the men was," she writes, "accompanied by a process of housewifization of women."[64] That meant not just kicking women out of the artisanal workforce, sabotaging their control of their fertility, and legalizing and then more or less requiring working-class marriage (and its dominating husbands) but also carrying out a lethal reign of state terror against women who refused to get with the program and demonizing them as witches. "Proletarian men do have a material interest in the domestication of their female class companions," Mies writes. "This material interest consists, on the one hand, in the man's claim to monopolize available wage-work, on the other, in the claim to have control over all money income in the family."[65] Even though male workers have been forced to cede a good bit of that territory since Mies wrote *Patriarchy*, in the 1980s, the sexual division remains. The persistence of a domestic sphere where housewifized work appears *natural* in contrast to wage labor—despite their common usefulness—devalues women as workers. Their work counts less of the time than men's work.

If public-power strategists want to assume final responsibility for social planning, that means confronting the question of housewifized labor. This is not a purely conceptual question: It has practical implications, such as, does public power mean wages for housework? And if so, how would public-power authorities deal with a housework strike? What would the role of husbands be in contract negotiations? Whatever the answer, the abolition of the de facto class division between working-class men and women is both a prerequisite to public power and contrary to the interests of most working-class men. Public power's self-evident appeal to many American working-class men starts to evaporate as they leave the workplace. As a reserve of cheap labor, women assume and thereby externalize social costs, sometimes from the energy system. For example, environmental historian Stefania Barca uses the phrase "nuclear housework" to describe the way that societies count on women to handle the health costs of nuclear energy production, both in the event of disaster and under normal conditions.[66] Radioactive waste makes people who live in the wrong places sick, and it's women who end up taking care of people who get sick. Many ideologically progressive heterosexual male workers will no doubt embrace the opportunity for domestic equality—as some surely have already, independent of a larger social plan—but if housewives were worth killing to secure in the first place, I don't imagine everyone will line up to vote against a patriarchal social metabolism.

The common dynamic is what Brazilian economist and sociologist Ruy Mauro Marini called superexploitation.[67] Though no two workers find themselves exploited in exactly the same way under capitalism, we've seen how qualitatively different rates of exploitation persist across borders, races, and genders. Since Marini assumed that workers in general barely make enough to get by (by the standards of their own social position), to do even worse than that means to have *less* than you need to get by. Capital digs into the superexploited worker's life to get more, leading to "premature exhaustion."[68] As Minqi Li notes, we can measure the unequal exchange relationship between the United States and China in terms of years of labor, in life.[69] If public power means workers' lives won't be considered more or less worth living based on these divisions, that means abolishing those divisions, or at least the relative advantages they yield. Building a political constituency for that effort is more complicated than calling out the name "public power."

"Universalism does not exist: it is a work in progress," writes French feminist Christine Delphy. "And its realisation requires a denunciation of false universalism: for the principal obstacles to accomplishing universalism are those who pretend that it exists already."[70] To assume political unity out of the working class's shared structural condition means abstracting workers out of their real differences, reducing all of

them to their respective places in the labor-capital class dichotomy.*
Not even capitalism does that, not really. Rather than intellectually abstracting our way to public-power political unity, we must forge it as part of a dynamic process, in the world's uneven actuality.

THE AFFIRMATION TRAP

I have examined a number of ways in which capital confounds the conflict between exploiters and the exploited by privileging some workers according to crosscutting identities and social formations. But what if workers come to oppose a public-power transition strategy *as workers*? That would certainly throw a wrench into the plan. We can't just assume that unity into existence. In reality, it's not hard to find groups of organized workers who oppose the public-power agenda: Go to any picket line; they'll be the ones with the guns and badges.

Police aren't the only union members with something to lose in a public-power transition. Workers share a structural interest in the abolition of a system of production based on their exploitation, but they also share a seemingly more immediate interest in maintaining access to their jobs and increasing their wages. Unions in the capitalist West tend to spend much more of their time, energy, and resources on building the power to bargain within capitalism than on building the power to abolish capitalism. Though these goals overlap when teachers bargain for the common good, political synergy isn't always ready-made. As geographer Kai Bosworth notes in his study *Pipeline Populism,* "Pipeline construction unions fight on behalf of their jobs and employers rather than a livable future for all."[71] Bosworth might have said *against* a livable future for all, because that's what it means to build fossil fuel pipelines at this moment in history. These workers

* For more, see Michael A. McCarthy and Mathieu Hikaru Desan, "The Problem of Class Abstractionism," *Sociological Theory* 41, no. 1 (February 2023).

and their representatives are not wrong that a public-power transition would plan them out of a job and destroy the capitalist social-metabolic conditions that yield—contrary to all other logic—a positive valuation of the pipelines they build. In our terms, it would make their lives as they currently exist not worth living. This is contrary to the goal of class struggle "viewed from the workers' perspective," as historian Moishe Postone describes it, which "involves constituting, maintaining, and improving their position and situation as members of a working class"—not destroying it and remaking it anew.[72] It's one thing to want public power; it's another thing to need your job.

I've already covered the international dynamics that have sent investment capital chasing cheap labor and cheap nature around the world, and those dynamics have turned a good job into a relatively enviable situation, one worth trying to protect. As necessities such as housing, health care, energy, food, childcare, eldercare, and education become harder to afford, more workers fall into the technical condition of superexploitation, even in the West. The Endnotes collective pointed out back in 2010 in an essay on housing that "we are witnessing the breakdown of the ability of the working class to reproduce itself on the level to which it has become accustomed."[73] Public-power optimists might point to high-profile union organizing efforts at bastions of class disorganization such as Starbucks and Amazon, but those efforts are notoriously difficult. Labor reporter Hamilton Nolan put it this way: "If you were a person who paid attention to labor news casually, you could see a sustained pattern of exciting things happening. But if you paid attention closely, you could see that reality was not shaping itself quite so neatly to expectation."[74] Even when it flexes its muscles—as it did in the UAW's precedent-setting strike and contract negotiation—labor has been juiced on a cocktail of public-private marketcraft, the price for which is a surrendering of the decisive planning role that public power sets out for the class. It is easier to get a union contract for a battery factory when the federal government has decided to derisk the private investment in the first place.

In this environment, one of organized labor's key demands is that capital continue to supply the working class with opportunities to be exploited. Capital's lips are in permanent pucker, ready to spit you out of your job, out of your home, out of the social metabolism, and onto the street. Then the state will come by with a garbage truck and take away your tent under new anticamping laws that foreclose public access to supposedly public spaces, such as streets. The public workers who gather your stuff and thereby turn it into trash will belong to unions, and they will know damn well they have something to lose. Despite the low quality of jobs, "End my life!" has not yet made for an appealing political slogan for the global working class.*

Writer Joshua Clover calls this situation the "affirmation trap," in which "labor is locked into the position of affirming its own exploitation under the guise of survival."[75] Fossil fuel workers have a clear reason to worry, and it's not unreasonable for workers in general to fear what will happen to them in particular when society starts planning according to new values. What's to stop the new regime from planning to sacrifice me the way it might sacrifice the natural flow of a particular river for a hydropower project? The promise of welfare programs that will provide a base level of living for all people regardless of their work might not be believable or attractive enough. Better to fight for our needs *as* workers, even if that means protecting the exhausting

* "In the inclusion through exclusion that defines the labour norm today, labour finds no positive affirmation; once a pivot around which the labour class was organised as a united subject, it is now a dividing and defining line within struggles, between those who can find regular employment in the labour market and those who enter and exit the ranks of the reserve army. This class *belonging* is no longer a basis for unity, a cause for proletarians united in struggle. Competition to access the labour market and to keep a job is a matter of life and death for the worker." Katerina Nasioka, "The Proletariat Versus the Working Class: Shifts in Class Struggle in the Twenty-First Century," chap. 8 in *Open Marxism 4: Against a Closing World* (Pluto Press, 2020), 129–30.

system of production for production's sake. "Caught in the affirmation trap," writes Clover, "labor ceases to be the antithesis of capital."[76] That means labor also ceases to be the affirmation of public power.

The affirmation trap bears directly on the public-power green transition. It's an old stereotype that workers and environmentalists can't get along, but there's a historical kernel of truth to it. Left-wing environmentalists have taken to blaming themselves for the misunderstanding, partly to distance themselves from the environmental movement's capital-aligned factions. But that gives short shrift to a lot of those environmentalists, who *were* thinking about workers as much as they were thinking about the trees. Consider the classic conflict between tree-sitting environmentalists and loggers in the Pacific Northwest in the 1980s and '90s. The opposition is obvious, and loggers turned the northern spotted owl into a mascot for everything wrong with job-hating hippies.

But that's not how Earth First! organizer Judi Bari saw it. She saw herself and her comrades as being on the side of workers—workers who might want to go fishing or camping on occasion—as opposed to the clear-cutting, job-automating timber bosses. She told an interviewer at the time, "There are very few people who think the right of a few people to make $7.00 an hour for a few more years is worth the extinction of a 10,000 year-old eco-system. I'm one of the strongest advocates for the loggers and the millworkers. I'm certainly a stronger advocate for them than their own unions are, and I'm certainly a stronger advocate for them than their employers are."[77] As she pointed out, logging jobs were few and dangerous, and they involved chopping down so many trees in the name of Value that soon there wouldn't be any trees or logging jobs left.* Most timber workers aligned against EF! regardless,

* It seems that protection of the northern spotted owl did lead to logging-job losses. See Ann E. Ferris and Eyal G. Frank, "Labor Market Impacts of Land Protection: The Northern Spotted Owl," *Journal of Environmental Economics and Management* 109 (September 2021).

tying yellow ribbons around their front-yard trees to signal their allegiance to the bosses. Some workers went a lot further than that, and Bari's writings are thick with scenes of loggers punching Earth Firsters in the face in defense of their jobs. (They weren't the only ones getting hit—Bari recounts a sign from a timber-wife counterdemonstrator: "If you take my husband's job, he takes it out on me.")[78] Bari and EF! were right, of course, that no real analysis of workers' interests could conclude in favor of clearcutting old-growth forest, and no conceivable public-power regime would allow it—even the American federal government ultimately didn't—but the loggers followed their bosses' plans. By placing jobs first, these workers ended up affirming capital.

So far, organized labor has been an unreliable ally for public-power advocates, which is a big problem for public power. As one study of green-transition labor politics concluded, in a sort of empirical verification of the affirmation trap, "When they led and embraced transitions on their terms, California unions not only grew their own power but became potent movement partners advancing climate action with broader social gains. In contrast, when unions took a reactive or defensive transition strategy, taking direction from employers, they hindered or offered no help to movements struggling for transitions away from fossil energy."[79] An unwillingness to commit to confrontation meant ceding claims to the planning role, they found: "Union leaders' hesitation to challenge employers might not only limit their bargaining power but also constrain them to follow the direction of private capital investment."[80] In the Bay Area, labor unions have split over fossil fuel questions, such as a new tax on an oil refinery in Richmond and a coal export terminal near the Port of Oakland.[81] A study of energy-sector workers in Germany and South Africa concluded, similarly, that "unions neither naturally oppose nor enthusiastically support green transitions"—the ones that adopt conservative job-protecting orientations end up siding with the bosses against the environmentalists.[82]

Unions "with ideological orientations towards social justice rather than a narrow membership and business union orientation," however, tend to be supportive.[83] What this all suggests is that support for public power has to be built, not abstracted or assumed, even among literal public power workers.

That there's still work to be done is dramatically clear in New York State, where public-power advocates got their win with the Build Public Renewables Act. While strategists in the Democratic Socialists of America plan for the state's public power workers to grab hold of the planning function and trigger a worldwide cascade of social transformation, the local International Brotherhood of Electrical Workers is engaged in some of the worst planning you could possibly imagine, as a headline on the New York Focus website in June of 2022 proclaimed: ELECTRICAL WORKERS UNION FIGHTS TO EXPAND FOSSIL FUEL–POWERED CRYPTO MINING IN NEW YORK. Joined by the AFL-CIO, the IBEW lobbied Governor Kathy Hochul not to sign a bill (she did) that put a moratorium on new proof-of-work cryptocurrency mines powered by fossil fuels. Since cryptocurrencies produce zero social goods, there's no excuse for them except to appease capital. The IBEW's legislative council head, Addie Jenne, was clear: The crypto industry can "reignite" disinvested communities through jobs and taxes, Jenne told the Focus, and "provide buzz to attract a certain type of workforce... and attract cluster economic development."[84] It's not the first time the state's big unions have backed fossil fuels.* "We need a realistic approach to meet the state's bold climate goals and must leave every option on the table," said the New York State AFL-CIO president, Mario Cilento, in defense of gas-powered energy plants.[85]

Labor identity politics and affirmative wage struggles aren't going

* Nor are New York's workers peculiar; the State Building and Construction Trades Council of California has stubbornly supported the oil and gas industry. See Alex Nieves, "Labor Unions Are Still Giving Democrats Climate Headaches," *Politico,* December 4, 2023.

to do it, certainly not alone. Realism might be what Cilento's members need, but it's not good enough for the world they live in. To organize the social metabolism for its own well-being requires higher ambitions, the kind of principles you can't simply derive from having a lousy job. To remake the world better, we have to leave the workplace.

CHAPTER 3

COMMUNISM

I.

THE COMMUNE

Recall: Grigris. When we left our Chadian protagonist, he was smuggling petrol across the Cameroon border to secure the money he needs for his beloved stepfather's medical treatment. But though the dangerous work is better paid than a safer, more ecologically responsible job would be, it's still not enough. Pushed up against a wall, he steals a load of gas, pays for the treatment, and fakes a robbery. But Grigris's employer doesn't believe his story, and the law of Value says he must pay with his life. Theft can't break the Value-Life link; it can only change whose life is forfeit.

But Grigris doesn't sacrifice his life for his stepfather's. Instead, he and his pregnant girlfriend flee the city for an agricultural village where the only inhabitants seem to be women. The villagers accept the couple gladly, celebrating the addition to their community. When a smuggler henchman comes, gun in hand, to reconcile the books, the village women surround him, each holding a large stick. None of them hits the assassin very hard, not individually, but the blows accumulate fast, and soon they're burning his body along with the car he rode in on. Capitalist Value collided with the village's communal values, and, faced with a violent challenge to their plan, the villagers defended their own social-metabolic order. If capital insists that saving a life that's not worth saving is a violation of the law of Value and that the penalty is death, then the village asserts its values in a parallel way. Many sticks can not only beat one gun, they can also smash the capitalist chain.

This chapter is about the strategy of many sticks, even if the name has other associations. The term *communism* is weighed down by a lot of history, and it may seem like I already covered communism in the public-power chapter, which imagines organized workers taking control of the means of production. But there are important theoretical and practical differences between these second and third strategies, as many as between the first and the second, and I believe they will be easy enough for the reader to understand. *Communism* is the best term I could find to describe a strategy in which the planet's exploited people abolish capital's system of Value and impose a new world social metabolism based on the interconnected free association and well-being of all—and not just humans.*

"What do we need besides equality of rights?" asked an anonymous group of writers in the French Revolution's wake. "We need not only that equality of rights written into the Declaration of the Rights of Man and Citizen; we want it in our midst, under the roofs of our houses."[1] This 1796 "Manifesto of the Equals" called for the revolution to go beyond political equality. They wanted "the common good or the community of property": an end to private ownership of land, before the early capitalist class had really started getting good at it. There's a green

* And yet I did consider some of the euphemistic variations: *communalism,* or *commoning,* or maybe *communization,* if only to make the substance of my work in this section easier. But in the book so far I've worked hard to use only the best words I can, and I feel it would be a violation of my implicit promise to the reader to do so if I were to stop now. As a matter of authorial approach, I would much rather trust readers to suspend their judgment and allow me to explain what I mean than suspend my trust in them and start using unnecessary syllables.

ecological tinge to the document, too, with a vision of common land as the basis for collective thriving. "Families moaning in suffering," they wrote, "come sit at the common table set by nature for all its children."[2] A say in planning isn't good enough: Communists want to restore people's ability to reproduce themselves via direct metabolic connection with their particular environment, with the land. Not as individuals, of course—imagine, a commune of hermits—but, like all animals, as members of a species in relation with other species. Communists aren't afraid of sounding archaic, but as Mariátegui put it, the past is "a foundation, not a program." (This goes for idealistic dreams to reconstitute European peasant life as well, as if capitalist history could be rewound and paused.*) The past is a foundation as people make a *new* world: despite, against, and after capitalism.† As in the relatively happy ending of *Grigris,* communism and capitalism can't coexist without a war.

From the communist perspective, the capitalist world is split by metabolic rifts. In addition to being unsustainable on a planet with a relatively finite number of atoms, accumulative, exploitative, and extractive practices generate pathological inequality. Rifts emerge everywhere—between cities that throw away imported nutrients and rural areas that have to deal with their shit; between humans and the other animals we instrumentalize; between core countries and the periphery, mined for cheap minerals and cheap labor. "If production systems result in waste products that threaten biological reproduction and social problems that weaken the social whole, an ecological and social crisis occurs," writes the ecofeminist thinker Carolyn Merchant.

* On this question, see Alex Heffron and Kai Heron, "Let a Thousand Fiefdoms Bloom," *Spectre Journal,* November 25, 2020.

† "After all, the commune form *is* the primary age-old way of bringing people together through association and cooperation. Embracing the vitality and conviviality of an archaic form like the commune seems, if anything, to enhance the ability of groups and individuals to confront the new social, economic, and ecological conditions of the present." Kristin Ross, *The Commune Form: The Transformation of Everyday Life* (Verso Books, 2024), 127–28.

"A contradiction opens between production and reproduction."[3] The earth's systems—including but not limited to its human systems—are superexploited and can no longer sustain themselves. (This was apparent in 1996, when Merchant published her argument, but the ravine is deeper now.) She calls for an alternative ethic of "earthcare," one based on healing the rifts between humans and the rest of our world so that we might rejoin it as partners. And if humanity could shake off capitalism like water off a dog's back, we could adopt a more earthcareful social metabolism without communists having to beat anyone to death with sticks, so to speak. Let us pretend for a moment that this is the case.

Whereas the marketcraft and public-power plans rely on outcompeting the capitalist system from within, communists seek to overturn capitalism and abolish Value as a social mediation. "We cannot keep things the same and change everything," argues the communist writer Jasper Bernes. "We need a revolution, a break with capital and its killing compulsions.... A revolution that had as its aim the flourishing of all human life would certainly mean immediate decarbonization, a rapid decrease in energy use for those in the industrialized global north, no more cement, very little steel, almost no air travel, walkable human settlements, passive heating and cooling, a total transformation of agriculture, and a diminishment of animal pasture by an order of magnitude at least."[4] Those specifics don't define the communist agenda; rather, they are *implications* of the dual injunction to repair the rifts and build equality. The end of cheap nature and cheap labor is, without exaggeration, the end of the world as we know it, so it's not hard to point out particular changes that we could expect to follow. No one thinks this process will be automatic or easy or without heavy costs, just that it's necessary. For the communist, the end of this world is long overdue.

With the middle link shattered, the Oil-Value-Life chain can't hold, and we're forced to pick Life off the ground and figure out what else to do with it. We'll have to plan a world, but according to what? Without Value to make everything commensurable and generic, we're tossed into the big wide sensuous world, left to shield our eyes from the bright light of unmediated reality. What we need is an orienting principle—an algorithm to substitute for capital's "production for production's sake," a value to replace Value. The communist orienting principle is simple: From each according to their abilities, to each according to their needs. While capitalism dictates what's valuable but hesitates to describe anyone's needs, communism has to define some needs in order to make sure they're being satisfied. For example, communists have good reason to believe that people need nutritious food, so we must plan a world in which everyone has plenty of nutritious food. (Recall István Mészáros and his "primary mediations.") Capital, by contrast, recognizes a *market* for food—nutritious and mostly otherwise—and produces into it, but it has no compunction about planning a world in which many people throw food in the garbage and many other people don't have enough to eat. For a communist, this would be untenable, no less a violation of society's cherished orienting principles than unceasing bank robberies would be for a capitalist. The "nature" that "sets a full table for everyone" includes people, which makes wasting food while others starve an *unnatural* act. Many human beings already understand this intuitively, yet we find ourselves locked in an indifferent capitalist social metabolism that forces us to naturalize that kind of obviously abominable behavior in order to secure a modicum of psychic well-being—this is one of capitalism's minor crimes.

Just as Bernes draws specific implications from the communist abolition of cheap nature and cheap labor, "from each, to each"—excuse me for using the celebratory acronym FETE going forward—implies things for society to care about and things for it not to care about. Quantity matters less, and quality matters more. As Mészáros writes of communism, "Use or utilization is, after all, what really matters in the satisfaction of human need, not the exclusive legal entitlement to little used or unused possessions."[5] From our seat behind the fogged-up capitalist windshield, it's not precisely clear what the relationship between terms like *usefulness, efficiency,* and *elegance* might be in a communist world, but we can predict that such values will be more important to a social metabolism based on need than to one based on capitalist Value, which spits on that interrelated series of concepts. Rather than exchange their time for the commodities they need, communists exchange useful activities to constitute a joyous and reproductively successful social metabolism. It takes more than a hammer and nails to make a house into a home, and we have to be careful not to confuse useful with Valuable in the capitalist sense. The web of life is impossible to map, and there's no communist call to discount, say, an experimental artist's abilities just because we don't immediately understand them, not as long as everyone's needs are met. But how do communists decide if everyone's needs are met?

While marketcraft settles within the existing framework and public power seeks to shift democracy's remit from the inside, communism cannot tolerate the persistence of top-down politics, not even within its constituent cells. FETE requires individuals to figure out their own needs and abilities, and they can't do that at the command of a president or a congress or a boss or a chief or a husband or a father. In the world's full light, theirs can only be a partial, false discipline. That hardly forecloses our collective ability to organize ourselves, however. Humans have experimented with many egalitarian tools of political organization, such as councils and public meetings and committees and spokespeople and facilitators and plebiscites and sortition. These forms—and plenty of others—help people concert their efforts without creating a technocratic coterie delegated to think and decide on everyone's behalf. There needn't be any overall decision-making power; not even the brain is president of the body.

There's a danger of communism devolving into localism, which would involve rejecting the best thing about the long twentieth century: the emergence of a true world society. That's not only unworkable, it's also deeply undesirable. We have to figure out how to collect ourselves as a species at all levels, from the local to the global and everything in between. "How can we preserve a ground, a home, a self, while being part of a whole where this self merges into an abstract we? How can the Earth," asks researcher Malcom Ferdinand, "be made into a *world-ship*?"[6] To universalize particularity is to cultivate forms of social organization based on people's actual intercourse with the world, which always happens in (across, between, and so on) particular places. The cellular structure for communism is the commune: a group of people who make FETE a reality together, since no one can do it alone. Just like human cells, communes constitute atomic wholes, and yet they only ever exist in relation to one another, in the context of a larger body. It does not make sense to say the heart is more important than a cardiac cell; they're the same thing at different levels of magnification.

The cellular social-metabolic unit of the commune marks a big difference from previous strategies, which still use the family household as their base unit, preserving the so-called private realm of reproduction. Since communism involves stitching the rift between production and reproduction back together in the earthcareful way Carolyn Merchant describes, it has to bust the division between private and public and revolutionize all parts of social life, even the most intimate. It's no wonder many people have found the idea haunting, despite state communism's eclipse over the course of the twentieth century. Just as the boss always has to be afraid of the worker, the patriarch knows the communists are ever-present, waiting to knock down the door to his home.

What do communists mean when they say they want to abolish the family and replace it with the commune? M. E. O'Brien unfolds the answer from the question of freedom as it relates to gender and sexuality, which are traditionally filed under "private" and not subject to questions of public planning. Under communism, she writes,

> Sexual and gender freedom necessarily means that how people choose to organize their romantic lives, kinship networks and domestic arrangements should have no consequence for people's standard of living and material well being. Gender freedom, therefore, relies on the widespread accessibility of means of survival and reproduction that do not rely on the family, wage labor, or the state. These means of survival include both the material features of reproduction—housing, food, hygiene, education—and the affective, interpersonal bonds of love and care people now primarily meet through family. Care under communism could be a crucial dimension of human freedom: care of mutual love and support; care of the positive labor of raising children and caring for the ill; care of erotic connection and pleasure; care of aiding each other in fulfilling the vast possibilities of humanity, expressed in countless ways, including through the forms of self-expression now called gender.[7]

Family abolition does not mean you are forbidden to hang out with the people whose genetic material combined to form you, but it *does* mean you have plenty of other options if that doesn't work out, even if you're just a kid and can't wield any public power. FETE includes a consideration of hugs because people do need hugs. This radical feminist demand of universal in/dependence is at the core of what it means to build the communes, and it's no coincidence that, around the world, women have been at the twenty-first-century communist vanguard.

NEW ANARCHISTS AND OLD COMMUNISTS

The reader might be surprised to hear that there is a twenty-first-century communist vanguard at all. Public-power socialists, sure, but *communists*? Not all the movements grouped in this chapter identify with the label; maybe most of them wouldn't. And yet in the course of historical events, they've acquired a common communist character. Whereas communists may be hard to find, communes are popping up all over.

In early 2002, anarchist anthropologist David Graeber published an essay in the venerable anglophone socialist journal *New Left Review* titled "The New Anarchists." Graeber pulled no punches, addressing the publication's contributors as much as its readers: "It's hard to think of another time when there has been such a gulf between intellectuals and activists; between theorists of revolution and its practitioners," he opened. "Writers who for years have been publishing essays that sound like position papers for vast social movements that do not in fact exist seem seized with confusion or worse, dismissive contempt, now that real ones are everywhere emerging."[8] Rather than the industrial proletariat's foreordained rise, Graeber pointed to what he called (against the grain) the "globalization movement"—a network of communities coming together to both defend their communal ways of life *and* combine them into a world-ship for the new century. Graeber described the admirably variegated crew: "Not only anarchist groups and radical trade unions in Spain, Britain and Germany, but a Gandhian socialist farmers' league in India (the KRRS), associations of Indonesian and Sri Lankan fisherfolk, the Argentinian teachers' union, indigenous groups such as the Maori of New Zealand and Kuna of Ecuador, the Brazilian Landless Workers' Movement, a network made up of communities founded by escaped slaves in South and Central America—and any number of others."[9] At the front were the Zapatistas, a pseudomilitary (but antimilitarist) movement emerging from the Lacandon jungle in the Mexican state of Chiapas in the mid-'90s.

Though the Zapatistas—Ejército Zapatista de Liberación Nacional, or EZLN—have defied simple categorization within the left wing, they fit the communist category as I've defined it just fine. Maybe it's better to say they shaped the category itself. The EZLN is a historical product of an encounter between some survivors of the country's Cold War anticommunist purges and Indigenous communities in the southern border region. Mexico's working class suffered in the '90s as a capitalist offensive trampled the country's public sector in its hunt for cheap labor and cheap nature, pushing Indigenous workers into relatively unproductive parcels of collective agricultural land in the rainforest. The North American Free Trade Agreement (NAFTA) generally stands for this whole set of shifts, but it was preceded by a shredding of the country's revolutionary constitution and the breakdown of the International Coffee Agreement in 1989—the resulting price collapse hit Chiapas especially hard.[10] The Zapatista uprising began with NAFTA in 1994, taking a forceful but confusingly nonviolent form. When the EZLN would take a town, they'd break people out of jails and destroy property ownership records, acting more like classic peasants than Marxist guerrillas, despite the balaclavas. After two years of struggle, the Zapatistas signed the San Andrés Accords, which drew on the International Labour Organization definition of Indigeneity and recognized Indigenous rights in the Americas for the first time since European colonization.

The San Andrés Accords not only (officially) guaranteed Indigenous equality within the country's legal structures—ceding public power to those who'd been held outside the political public—it also recognized Indigenous autonomy, the ability to "decide upon their form of internal government and their ways of organizing themselves politically, socially, economically and culturally," and made ecological sustainability one of the bases for a new relationship between the state and its tribally organized peoples.[11] (Once again, we can recognize the shape of "social metabolism" in a constellation of other words.) Though their terms have been hard to enforce, as I'll explain, the accords showed that in the face of newly intense capitalist encroachment, Indigenous people had a viable choice other than resignation. In 2020, a group of scholars published an article in the peer-reviewed geography journal *Geoforum* that looked at the Zapatista uprising from an interesting angle: its impact on agrobiodiversity. The general assumption is that armed conflict is bad—"not healthy for children and other living things," as my grandmother's poster had it—but these researchers couldn't help but conclude that, judging by the EZLN's measurable success in improving community food sovereignty, saving heritage seeds, and adapting to climate change, "conflicts may also create favorable political conditions for the implementation of community-driven agrobiodiversity management."[12] What they mean but don't quite say is that it matters who wins.*

The *Geoforum* scholars end up describing the Zapatista agroecological communes as something parallel to public power's unified subject-object: "Agrobiodiversity in these communities has become not only a means for building autonomy, as captured in the Zapatista slogan 'without food, there is no resistance,' but also an end, in the EZLN's decades-long effort to improve the material and environmental conditions in which its autonomous communities live and reproduce their

* Other scholars have drawn similar conclusions about deforestation in Colombia: "Our results show that, within a municipality, a higher [right-wing] paramilitary presence is associated with higher deforestation rates, while the opposite occurs for [left-wing] guerrilla groups. We believe that these contrasting impacts are a result of paramilitary and guerrilla differences in ideology and modus-operandi." Tatiana Cantillo and Nestor Garza, "Armed Conflict, Institutions and Deforestation: A Dynamic Spatiotemporal Analysis of Colombia 2000–2018," *World Development* 160 (December 2022).

subsistence agriculture."[13] In his essay, Graeber references the Japanese peasant resistance against expansion of the Tokyo airport as a precedent. In a 1973 documentary detailing the fight from the villagers' perspective, the Ogawa Productions collective relays the story that when pro-airport families left the community, everyone else came together to farm their plots and use the proceeds for antiairport activities, including bailing out members of the Young Farmers Action Group.[14] The authorities called this theft and trespassing, but abandoning the fields was contrary to the village's ecosocial metabolism, contrary to nature: "The sight of these wasted fields," one villager recounts, "repelled the farmers like their own flesh rotting."[15] In one village meeting scene, they decide that at harvest time they will all inform on their neighbors' needs to better coordinate help—villagers were being too modest about asking for assistance, and FETE means FETE. Together, they formed a communal fighting apparatus, all the stronger through its dedication to care. Capital weaponizes care by withholding it; communes weaponize care by providing it as freely as possible.

Graeber suggests that "the extraordinary importance of indigenous people's struggles in the new movement" might be explained by their contradictory position as the most oppressed and yet least alienated people in the capitalist world.[16] That explanation may hold some truth, but I fear it bleeds into psychologism that is beyond my ability to evaluate. Instead, I look to the same place Mariátegui did: land. In the opening essay to their 2005 collection, *Reclaiming the Land: The Resurgence of Rural Movements in Africa, Asia and Latin America,* editors Sam Moyo and Paris Yeros of the Sam Moyo African Institute of Agrarian Studies write that "capitalism has subordinated agriculture to its logic worldwide, but without creating... home markets capable of sustaining industrialization, or fulfilling the sovereignty of decolonized states."[17] The authors call this situation semiproletarianization: "Neither full peasants nor settled proletarians, semi-proletarians have grievances that arise from both the family farm (land shortage, insecurity of tenure) and the workplace (wages and conditions of employment)."[18] Superexploited, semiproletarians rely on extra-market (family) reproductive labor, small garden plots, and gray-market precarious employment in order to get from one day to the next. Cheap nature makes work for semiproletarians—illegal logging, petrol smuggling, artisanal mining, and so on—which worsens metabolic rifts and makes subsistence harder. As capital threatens tribal and peasant communities with semiproletarianization, Moyo and Yeros point to "the incipient indigenization of Marxism."[19] It's a prediction that's been borne out in the decades since the book's publication, as Indigenous land and water defenders have taken the lead in challenging capital around the world and Indigenous theorists have pushed the Marxist tradition in fruitful new directions.

Part of this Indigenization of Marxism is the reclamation of an Indigenous place in the Eurocentric history of resistance to capitalism and alternative models of communal life. "There is an assumption that socialism and communism are white and that Indigenous peoples don't have this kind of thinking," writes Nishnaabeg theorist Leanne Betasamosake Simpson. "To me, the opposite is true. Watching hunters and ricers harvest and live is the epitome of not just anticapitalism but societies where consent, empathy, caring, sharing, and individual self-determination are centered."[20] This quotation comes from Simpson's book *As We Have Always Done,* and she weaves Nishnaabeg stories into the fabric of anglophone anticapitalism. She writes of the mythological character Nanabush, who tries to live by exploiting cheap labor and cheap nature but always learns his lesson. She writes, too, of his brothers, who game the FETE system and consume without contributing.[21] One becomes a celebrated artist—revealing that not all contributions are immediately apparent as such—while the other refuses to engage in any communal practice at all until he shrivels up and becomes moss. These stories crystallize Nishnaabeg values, and Simpson poses them as more than just important cultural objects: They are forms of communist *planning* at the social-metabolic level. Simpson describes Nishnaabeg social metabolism as distinct from public power: "We didn't just control our means of production, we lived embedded in a network of humans and nonhumans that were made up of only producers."[22] To say everyone is a producer, from the smallest microbe to the most powerful human engineer, is to explode the unproductive/productive binary and recognize what it is we're all building together: the world-ship.

The formulations here may sound a bit loose. FETE is not a law that can be imposed or a right that can be asserted. But by engaging with contemporary Indigenous theory, communism finds some answers. The Indigenous legal scholar Val Napoleon argues convincingly that this "is not about trying to go back in time, but about drawing on the strengths and principles of the past to deal with modern-day problems and situations."[23] (We have plenty of those.) Once again, the Indigenous past is a foundation, not a program. "Law is one of the ways we govern ourselves," she writes. "It is law that enables large groups of people to manage themselves. Law is something that people actually do. Indigenous peoples applied law to harvesting fish and game, the access and distribution of berries, the management of rivers, and the management of all other aspects of political, economic, and social life."[24] At this level of abstraction, law reveals itself as ecosocial-metabolic planning, not a book of rules. And since Indigenous laws relate to and are grounded in particular ecosystems, there's only so much we can universalize. That's a good sign, given that generic visions of the earth have rendered it decreasingly livable. FETE only works as a product of harmony between a community of humans and their specific environment—the water, weather, plants, animals, and so on. However, these are dynamic systems, and tradition without the renewing power of critical thought threatens to degenerate and become arbitrary. Because these laws are stored in names and traditions and practices, Napoleon writes, "sometimes it takes more thinking to see the usefulness and practical application of these traditions—or, for that matter, their uselessness." That task is perfectly suited for the Marxist practice of "ruthless criticism of everything that exists" (RCEE), a lawlike rule that applies soundly in all particularities. RCEE is a second universal law to go with FETE: contemporary contributions from an ancient wandering tribe that have been refined through centuries of practice, Indigenous laws of the literate but landless. With those powers combined, the world-ship steers, and we can imagine a captained planet.

But perhaps we should be wary of looking for a communist captain or even a communist captain species. Diné–Laguna Pueblo artist Steven Yazzie offers a different metaphor in his 2003 painting *Orchestrating a Blooming Desert*. It shows the back of a man dressed in blue jeans and a suit jacket as he faces a desert landscape. Rather than the typical red and orange, this desert is bursting with green: hills full of grasses and shrubs and cacti. In the man's left hand perches a small bird; in his right is a conductor's baton. The foreground is dotted with a variety of stemless floating flowers that swirl around him like magic. Maybe this is better—humanity as the conductor species, drawing out harmonies from a landscape instead of trying to chart its path.

AGROECOLOGY

By what criteria do Indigenous communal systems outperform the capitalist social metabolism? What makes them more harmonious? After all, capitalists aren't the first human planners to shred their environmental systems—whole civilizations have plunged themselves into historical darkness thanks to poor social planning and resource management and/or failure to adapt to changing ecological circumstances.

Under the harsh light of RCEE, who's to say that enduring Indigenous laws are any better than production for production's sake? Can we even speak of Indigenous laws as such, given the wide variety of systems around the world? But if the communist plan is to repair the rifts capitalism has torn, there are sound reasons for turning to Indigenous answers.

In 2021, the Food and Agriculture Organization of the United Nations issued a 420-page report on Indigenous food systems. The report profiles eight groups around the world and draws some common conclusions:

> The territorial management practices of Indigenous Peoples are carefully attuned to the ecosystems in which they live and ha[ve] been able to successfully preserve biodiversity and create sophisticated food systems that generate food for communities for generations. Scientists are starting to acknowledge this whilst policymakers have not yet been able to translate this growing awareness into effective policy measures that protect Indigenous Peoples' practices. There is potential to draw lessons on sustainability from Indigenous Peoples that can be extrapolated to other contexts and communities.[25]

While capitalist policymakers ineptly continue to degrade the earth, Indigenous policymakers are faring much better. Though there are approximately half a billion Indigenous people in the world—somewhere between 5 and 10 percent of the global population—scientists commonly credit their managed territories with maintaining 80 percent of world biodiversity. An oft-cited analysis published in the academic journal *Nature Sustainability* calculated that Indigenous peoples manage 37 percent of remaining natural lands. "Indigenous Peoples often manage their lands in ways that are compatible with, and often actively support, biodiversity conservation," the authors conclude. "They can co-produce, sustain and protect genetic, species and ecosystem diversity all over the world by 'accompanying' natural processes, for example creating cultural landscapes with high habitat heterogeneity and developing and restoring ecosystems with novel species combinations of wild and domesticated species. Furthermore, Indigenous-led approaches have highlighted innovative ways to design conservation reserves, environmental policy instruments, wildlife monitoring and management programmes."[26] What these researchers are describing in their own language is earthcare.

One reason to start with food when we look at land management and planning is that we all need it. Here is ecosocial metabolism at its most literal: People must digest our environment in order to live. To re-create the exploited population as waged workers, capitalists fenced in the world and made it illegal for people to reproduce themselves off the land directly without an official title. Instead of peasants paying their lords with a rent share of produce and a periodic term of forced labor (corvée), the planters paid the farmers a wage with which the farmers could buy food. Though this original accumulation—in which some people enclose land, start the capitalist game clock, and claim legal ownership—is partly periodizing, capital is also always starting anew. As unpermitted cattle ranchers push into the Amazon rainforest and Israeli settlers pull up olive groves in the West Bank, they're performing original accumulation even as I write, in the 2020s.

If capital is advancing, then there has to be somewhere for it to advance on, a frontier. As I've said, Indigenous peoples around the world maintain some level of nonmarket communal reproduction. According to the UN report, all eight surveyed Indigenous groups get most of their food from nonmarket sources. The Baka people of the Congo basin rainforest, for example, produce more than 80 percent of their food directly from 179 plant and animal species (not counting three stimulant plants). "The Baka depend on the forest. We use it to eat, to heal ourselves and even for our traditional rites," a young Baka man named Tom told a reporter for *Le Monde* in 2022. "But we can no longer access it."[27] The Baka's forestland has been enclosed by a strange kind of capital: conservation philanthropy. Western donors are getting tax breaks as they pay armed guards to force some of the world's most effective biodiversity managers off their land, all in the name of saving animals.

If capital insists on estranging humans from their environments, then communists vow to repair that rift as well. As Moyo and Yeros document, the attacks on Indigenous and peasant land management have yielded illegal land occupations as the technically displaced refuse to leave. The northeastern Brazilian coastal state of Bahia has a long tradition of resolute Indigenous self-management and Maroon communities of escaped Africans called *quilombos,* and since the 1990s, their land occupations have retaken territory from extractive capitalists. Launched in 2012 and inspired by the Zapatistas, the network Teia dos Povos (TdP) has attempted to coordinate these struggles with help from the local Movimento dos Trabalhadores Rurais Sem Terra (MST). More than just taking space, TdP is thinking about what kind of ecosocial metabolism to set up once they've got it. "We are wondering what to do beyond the fence. How to build territory," says TdP spokesperson Erahsto Felício. They build up a store of food, so the original gardeners can meet their needs as they work, then they cut the fence:

> We need to make the first garden to produce an amount sufficient for our existence: beans, sweet potatoes, and corn (three months), cassava (six months), and banana (one year) are species that we always plant because of their ability to generate a healthy and rich diet.

> While we are making this garden, we need to work hard to raise our agroforestry. For this, we have a seedling nursery with native and fruit trees (acai, cupuacu, cocoa, etc.). Growing the forest is the beginning of the land regeneration process.[28]

From there, more food supports more land occupations. Theirs is an earthcare insurrectionary strategy, slicing through the fence to stitch up the land. That's the kind of class war that, as Stefania Barca writes, is based on "a common material interest in keeping the world alive" among workers. To do that, we need to go beyond public power–style "calls to fully embrace the forces of production" and use those "sciences and technologies which are appropriate to, or already mobilized in, counter-master projects of earthcare."[29] We don't usually talk about seeds as technology—not unless someone spent a lot of money to toy with the genes. But for communists, the material ways in which humans plan their relationships with plants and animals and other earth systems is the most important kind of productive technology there is.* That's why it's common to find communists from the Amazon to Kurdistan to Japan talking about agroecology.

"Agroecology is the technological keystone," writes Max Ajl of the radical green transition. "The term refers to applying scientific experimentation to and the formalization of the processes underlying traditional farming systems."[30] One way to think about agroecology is as a great encounter between the RCEE law and the world's many varied Indigenous ecosocial metabolisms in the context of a sinking world-ship. What appear to be low-tech methods—Ajl cites polycropping, cover crops, vertical forest gardens, raised beds, Vietnamese fish-garden-pork systems, and more—are earthcareful practices that

* For a communist account that attempts to transcend the nature/technology binary when it comes to growing food, see Out of the Woods Collective, "Contemporary Agriculture: Climate, Capital, and Cyborg Agroecology," in *Hope Against Hope: Writings on Ecological Crisis* (Common Notions, 2020), 105–18.

reduce the need for external inputs and human labor. The challenge, he writes, is finding contemporary analogues for these practices that are equally efficient in terms of labor and energy but can scale up to support much denser populations.[31] Learning a lot more about the world, both in its generalities and its particularities, is the only way for us humans to find sustainable roles on the planet, and we better do it fast.

Critical or communist agroecology calls for more than the revival of traditional practices; it's a plan to plan the world for the world. That means Graeber was correct to call this movement the globalization movement: The collaboration of agricultural communities across borders has already outlived the short period in which the world's capitalists united under the banner of free trade. Though a term such as *food sovereignty* might sound reactionary and autarkic, this idea—that any community should be able to feed itself regardless of its position in the international market—emerges from La Via Campesina, a global network of peasant and Indigenous agricultural organizations formed in the 1990s in opposition to capital's reworking of the value chains and the resulting semiproletarianization. Agroecology is a planet-scale science and thus an organizing practice as well. Consider Nous Sommes la Solution (NSS), an international network of hundreds of West African rural women's associations devoted to asserting food sovereignty, building power, and spreading agroecology under the ecosocial-metabolic slogan *Produisons ce que nous consommons et consommons ce que nous produisons*—Let us produce what we consume and consume what we produce.* In 2020, the Senegalese unit of NSS shared an example of what that looks like with their Sum Pak project. Analyzing changes to local traditional diets, NSS members focused on the prevalent use of industrially produced bouillon cubes, which make cooking stews and rice dishes much easier but also contain unhealthful amounts of salt,

* Compare the group's name to one of Karl Marx's definitions of communism: "Communism is the riddle of history solved, and it knows itself to be this solution."

leading to novel health problems. These cubes also undermine local agriculture and community food sovereignty, making agriculturalists dependent on capitalist factories in order to feed themselves. Not to mention the fact that, as NSS president Mariama Sonko said in her presentation on sum pak, the bouillon cube companies throw in racist, sexist advertising rhetoric for free.[32] NSS in Senegal put these insights into practice by designing, producing, and circulating its own agro-ecological bouillon product—sum pak, made from local ingredients such as shrimp and locust beans—and sharing the results through the larger network. The goal of agroecology is hardly to eschew technology—NSS calls for the wider distribution of solar-energy infrastructure, for example—but to subject it to RCEE by communist rather than capitalist values.* Concentrated flavor powder is a great idea, but what's inside the cube and how it gets there matter.

In the globalization movement, agroecological techniques can jump continents without having to go through capitalist control. In their book, *All We Want Is the Earth: Land, Labour and Movements Beyond Environmentalism,* scholars Patrick Bresnihan and Naomi Millner describe the "*campesino-a-campesino*" transmission of knowledge at the end of the twentieth century and into the twenty-first. They explain how bokashi, a method of making natural fertilizer, traveled from Japan to the mountains of rural El Salvador via Costa Rica. Bokashi—codified by agricultural scientist Teruo Higa in the early 1980s—relies on active fermentation, requiring agriculturalists to add supplemental microorganisms directly to waste. "As agroecological farmers have adapted and shared bokashi, so its recipes have multiplied, reflecting local availability but also the variability of traditional practices," Bresnihan and Millner write. "Usually, the process involves arranging, in

---

* In another example of the agroecological use of RCEE, the sum pak project decided to replace sand—traditionally used to shell locust beans for paste—with the more appetizing rice bran.

a covered area, solid materials (such as bran, rice husk, soil, manure, charcoal and ash) into a pile, together with a source of easily assimilated carbon (for example, molasses, honey or sugar) and a mixture of microorganisms dissolved in water. Yoghurt or yeast is used to multiply the effectiveness of the microorganisms."[33] Not only is bokashi far superior to fossil fuel fertilizers from an ecosocial-metabolic perspective, it's also available at little cost. It's easy to see why no capitalist would be particularly interested in spreading this technology and equally easy to see why communists would.*

So far, I've made agroecology sound like a wholly rural program, and though agriculturalist producers have led the movement, one metabolic rift that communism is supposed to heal is between food-producing, waste-consuming areas and waste-producing, food-consuming areas. Some of the most important agroecological organizing has centered on bridging the gap between urban and rural communities. "When the *campesino* and the *barrio dweller* look each other in the eye, when they hear each other's stories, class-based solidarity emerges," says Ana Dávila of Pueblo a Pueblo, a Venezuelan planning structure that brings together rural producers and urban consumers.[34] Founded in 2015, six years after the country introduced a legal structure for communes, Pueblo a Pueblo connects communes on both sides: Poor families in Caracas determine their collective agricultural needs in advance and share them with the collectivized growers, who can then avoid the open market and capitalist profiteers in their entirety. The relationship is still mediated by money, but prices are "transparent," which means the urban communes know what all the inputs cost because they plan production together. A "double participation ladder" models five back-and-forth movements between the two communities, leading

* For more on international *campesino-a-campesino* relations, see Miguel A. Altieri and Victor Manuel Toledo, "The Agroecological Revolution in Latin America: Rescuing Nature, Ensuring Food Sovereignty and Empowering Peasants," *Journal of Peasant Studies* 38, no. 3 (July 2011).

from the identification of campesino areas up and across to the storage and distribution of food via analysis, assembly, agroecological planting, and education. This ladder is designed to replace generic Value with specific ecosocial-metabolic planning practices. As organizer Gabriel Gil puts it, "A solidarious, fraternal, class-based connection emerges between the producer and the consumer. This encourages the *campesino* to produce with more care, with a lower toxic load, while the city dweller overcomes the condition of being an alienated consumer and may even come to Carache to help with the crop."[35] The goal of such a program is to bring the two sides closer together and ultimately heal the rift between them.

US economic sanctions on Venezuela have pushed agroecology further, as producers have had to replace imported inputs, from seeds to pesticides, with alternatives. For example, Pueblo a Pueblo developed a simple pesticide made locally from copper sulfate and lime, which not only reduced harm to pollinators but also cut costs by more than 90 percent. And rather than import seed potatoes from Canada, they've instituted a project called *Papa para la vida, no para el capital* (Potatoes for life, not capital) to raise native varieties for seeds and consumption. Here, too, agroecology is a contemporary critical scientific practice. As farmer Bernabé Torres describes it, "Some ask me: why do you grow native potatoes? They wonder if it is some sort of attachment to old traditions. To that, I say, no, we grow these potatoes because they are sturdy, they do well in our harsh climate, they are more resistant to pests, and they don't need agrochemicals."[36] They're also more nutritious and keep for a longer time, and they don't just do well in the soil: They also do well *for* the soil, helping restore its depleted quality.

With all this talk of science and productivity, we can't lose holistic sight of our object. A communist ecosocial metabolism is about more than the instrumental production and circulation of calories and about more than the measurable improvement of soil health and food nutrient density. In Gavidia, Venezuela, the local Native Potato Project sponsors an annual three-day ecofestival in December, which features educational workshops, poetry, plays, traditional songs, and, of course, potatoes. In communism, all parts of life are planned for in common, not just the box of stuff labeled "public." An anonymous group in Minnesota showed just how this is done by intervening in a local debate over a golf course. Starting in 1923, city planners dredged Rice Lake—a sacred Dakota marsh—to construct the Hiawatha Golf Course. Almost one hundred years later, after decades of recurrent flooding, the Park and Recreation Board decided to give up on the eighteen-hole course and plan something different for the space. Among the agroecological suggestions were proposals to turn the area into a regenerative "food forest." But the authors of the pamphlet "Make the Golf Course a Public Sex Forest!" had a slightly different vision: "We think this last idea is amazing; our only objection is that it doesn't go far enough. We don't just want a food forest. We want a sex forest," they wrote.*[37] The authors detail the historical suppression of the city's queer cruising scene and make a strong case that "the city is subsidizing a handful of mainly upper-middle class golfers while thousands of perverts have no good place to fuck." That, too, is an argument about social planning, about land use, and about people's needs.

GETTING WET

Previously, I asked the reader to assume that communists were free to implement their plans. From there, based on what I've laid out in this chapter, the general idea doesn't seem very complicated. Knowledge-intensive agriculture—and knowledge-intensive housing, and knowledge-intensive transportation, and knowledge-intensive energy, and more—require different kinds of work, and no one is

* "But with food too," they added in a footnote.

saying the transition will be easy, but they don't require *more* work than capitalism. We switch, not instantly, but through iteration and consultation and education and cooperation as well as through initiative and conflict and excitement and courage. The communist plan is to create an ecosocial metabolism that gives people a lot more personal control over their time, energy, and attention. And better than that, the control isn't the flimsy, nonrenewable type that's mined from the earth or stolen from the future. It's the kind you build and pass on to future generations. Who wouldn't want to study and practice the science of being a good ancestor? Even if not everyone wants to be a communist, enough people do in order to communize a lot more territory than is being held now, enough to provide an exit from Oil-Value-Life for the many people who need one. What's stopping such a movement from taking the world by storm? Capitalism's armed guards, of course.

Rather than a weakness of the communist strategy, the fact that communism finds itself in direct confrontation with police, border guards, capitalist private security, and the occasional state army is a strength. Where other strategies might see these elements as necessary projections of the democratic will or just workers stuck in bad jobs, communists recognize them correctly as enemies of the planet and its people(s). They are the ones who man the fences between people and the land, who enforce semiproletarianization in the periphery and guard migration entrances to the core. It's a combination of policies that activist Harsha Walia calls border imperialism, which she describes as "the role of Western imperialism in dispossessing communities in order to secure land and resources for state and capitalist interests, as well as the deliberately limited inclusion of migrant bodies into Western states through processes of criminalization and racialization that justify the commodification of their labor."[38] The truth of human equality strains against these fetters, and holding them in place takes not just fences and work but also violence.

The West has fortified its borders, and island points of contact between its core and the periphery become points of conflict: the Greek archipelago, the Canary Islands. The United States, the UK, and the EU have done their best to externalize their border problems to Mexico, Rwanda, and Turkey, respectively, and everyone has been building walls. The Migration Policy Institute has tracked a "rapid proliferation" of border walls around the world, from only a dozen at the end of the Cold War to seventy-four as of 2022, with another fifteen in development.[39] "No continent has been spared from the reinforcement and fortification of borders," writes researcher Élisabeth Vallet, "which has come to define the beginning of the 21st century."[40] To man those fences, states have increased their border and sea patrols. But the truth is persistent, and wherever there's a border, people will find their way through it. Border imperialism has hardly dented a rising human demand for freedom of movement, and after reviewing the data, demographer Joseph Chamie recently concluded that "it appears that governments are unlikely to be able to resolve the illegal immigration dilemma any time soon. In fact, the dilemma is likely to be exacerbated by increasing illegal immigration due to growing populations, worsening living conditions, and the effects of climate change in migrant sending countries."[41] But for communists, people moving to more habitable climates isn't a dilemma: It's a self-evident solution. The point is to help.

In the context of border imperialism, the act of reaching across to take another person's hand acquires a communist character. People who migrate in search of a better life—whatever their mix of excitement and despair—assert the universal truth of human equality against nationalism's fence fictions. People on the other side who are there to help a stranger likewise affirm their own true humanity. This is a dynamic that scholar Paul Gilroy calls "offshore humanism" in his description of a Greek man named Antonis Deligiorgis who dived into the sea to help a boat full of drowning migrants. Gilroy writes that Deligiorgis and the people he saved were all soaking wet, and their shared salty saturation "communicates something of the way that being human is transformed when the solidity of territory is left behind. We are afforded a glimpse of vulnerable, offshore humanity that might, in turn, yield an *offshore* humanism."[42] Laws that criminalize assisting migrants soak allies on the border's core side with criminality, whether offshore or onshore. Members of the group No More Deaths, who leave water and food for people crossing the Sonoran Desert, have faced prosecution and up to ten years in prison for their work.[43]

In Greece, where the economic and geographical distance between the core and periphery reaches its contemporary nadir, the movement is its most communist. "Tens of thousands of refugees from Syria, Afghanistan, and dozens of other countries, working closely with the anarchist and autonomous movements, squatted entire hotels and other abandoned buildings in Athens, Thessaloniki, and other cities in order to attain dignified housing," writes activist Peter Gelderloos. "They also run cafeterias, language classes, clinics, child-care services, and other activities."[44] Though not as established—or antiestablishment—pro-migrant municipal networks have sprung up: No Border network in Eastern Europe, Re.Co.Sol in Italy, Red de Municipios de Acogida de Refugiados and Ciudades Refugio in Spain, Association Nationale des Villes et Territoires Accueillants in France, the New Sanctuary Movement in the United States, Anti Raids Network in the United Kingdom, and more.[45] These urban land occupations in the capitalist core haven't been nearly as successful as their rural inspirations, though that hasn't stopped communists from trying.

Gilroy's theory of offshore humanism reminds me of a 1988 mixed-media painting by Thornton Dial titled *Refugees Trying to Get to the United States*. Against a stormy painted background of gray and blood red, there's a raft of twigs and wire. On the raft, fused into a mass, are a bunch of people, their limbs made of plastic straws and their genderless faces molded from putty. For the faces, Dial alternates between three racializing colors—brown, yellow, and pink—but all gape with the same bottomless uncertainty that Dial pulls from their tiny gash-mouths. I've seen the piece interpreted as referring to Vietnamese immigration by boat and to refugees headed to the United States in general. But how, then, to understand the artist's choice to race nine out of the sixteen figures as unmistakably white, in that phenotypically pink way?[46] I don't think he's referring to the pilgrims. Rather, I see Dial depicting working Americans themselves as refugees trying to get to the United States, as if citizens, too, were at sea, trying to find the country that is supposed to lie underneath their feet. As we know, there are lines of exclusion within countries as well as between them, and those lines have guards, too.

After Minneapolis police officers murdered George Floyd, in the spring of 2020, an outraged population burned down the Minneapolis Police Department's Third Precinct headquarters and immediately set off a national uprising for the value of Black people's lives and against the police who devalue them. In cities across the country, demonstrators flooded the streets, targeting the police not just as their immediate opponents but also as enemies. Movement luminary Mariame Kaba published an op-ed in the *New York Times* titled "Yes, We Mean Literally Abolish the Police," in which she wrote about such an abolition as the sweeping aside of an obstacle to "a vision of a different society, built on cooperation instead of individualism, on mutual aid instead of self-preservation."[47] The abolitionist movement embodied that communist spirit from the beginning: Demonstrators took the initiative to freely supply one another with water, food, and COVID-19 prophylactics in order to prevent the spread of disease at demonstrations. Some went further, smashing the division between public and private in order to appropriate capitalist commodities (so-called looting).* Others went further still: In Minneapolis, a crew of people took advantage of the riots to communize an entire Sheraton hotel, offering 136 rooms for free to people who needed them.[48] It took the American military to repress the George Floyd uprising, and a similar conflict threatens to spark at any time.

In American cities where housing costs are high and availability is low people have continued to build tent encampments—the National Homelessness Law Center tracked a 1,342 percent increase in reported camps in parks and other public spaces nationally between 2007 and 2017—which the police routinely attack and destroy.[49] When communists use the slogan "All cops are bastards" (ACAB), they're talking about the structural role of police within capitalism and asserting that that's what matters. It's more or less the same logic police officers use when they say—as they often do—that they're not destroying a homeless encampment out of personal malice but out of professional duty. However, when we say that the police are "just doing their jobs," we should also recognize that they are *just* doing their jobs, which is to say they're not doing anything more than their jobs, including acting like human beings who can perceive and feel compelled to act by the desperation of other human beings. Similarly, police lack the capacity to recognize that climate protests are ultimately in the interest of public safety and instead find themselves clobbering fellow humans who are trying to direct everyone's attention to a currently elapsing global crisis.

The police are capital's impersonalized trigger finger, pulling on a gun pointed across borders, across property lines, across classes, and across races. They are the status quo's heavily armed militia, designed to insulate the ruling class from popular pressure, and

* See Vicky Osterweil, *In Defense of Looting: A Riotous History of Uncivil Action* (Bold Type Books, 2019).

thus they constitute the most direct obstacle to all progressive plans for the future. The police are there to shoot you if you start acting too reasonably or too human. Communists recognize that the police (and police-like authorities) volunteer to place themselves outside the human community, and communists plan to defend themselves and their values accordingly. The Zapatistas, for example, are not commonly understood as a movement of Indigenous women practicing agroecology—similar to the NSS—even though that's in large part what they are.[50] Instead, people know them as masked militants wielding AK-47s and marching in columns through the jungle. Communists are willing to do the latter to achieve the former without too much concern about whether their actions are legitimate in the state's eyes or not. As the communist poet and playwright Bertolt Brecht put it: "What is the use / Up to your chin in the shit, in keeping your / Fingernails clean?"[51]

ROSE THEORY

Living things tend to defend themselves from outside threats, just as they tend to breathe and eat and reproduce. Self-defense is part of humanity's individual metabolism and our social metabolism, just as it is for other animals. Kurdish political leader and revolutionary theorist Abdullah Öcalan has called this idea rose theory, based on the fact that even roses sprout thorns to defend themselves. "Do we not have the right to be protected as much as a rose?" he has asked.[52] Communists can and do assert their values in capitalism's cracks and interstices, but at a certain point two groups dispute control over a single piece of land according to different standards. If capitalists say it's their land because they have a fence, and communists say it's their land because they're caring for it, who's right? In most places and scenarios, capitalists have the law, but communists have the reality of a global environmental crisis. Between these dueling claims to legitimacy, who is the rock and who is the scissors?

According to rose theory, living things evolve species-specific defenses: Sharp stems survive. Humans don't have particularly sharp teeth or claws, but we can be thoughtful and crafty and inventive. Our most important defensive quality is not, as it sometimes seems to be, the refined ability to fling stuff super hard. Rather, it's our ability to reason, to imagine what isn't yet based on what is and act consciously with regard to events that haven't occurred. Communists have a good case that it's unreasonable and therefore unnatural to continue planning society according to Value when it's clear that will result in accelerating peril for the planet's creatures, including and especially humans. "How could we deem 'realistic' a project of modernization that has 'forgotten' for two centuries to anticipate the reactions of the terraqueous globe to human actions?" the philosopher Bruno Latour asks. "How could we call 'rationalist' an ideal of civilization guilty of a forecasting error so massive that it prevents parents from leaving an inhabited world to their children?"[53] If capitalism is an ecocide pact, humanity must back out; there can be no reasonable way to ruin everything. Just as the individual body pulls away from fire or revolts at drowning, society must defend itself from the climate crisis.

Where, then, is the line between self-defense and offense? There's no common agreement; it's a political question. One person's land defense is an interagency task force's ecoterrorism. By asserting a prerogative to perform earthcare in spite of capitalist property claims, communists throw the calculation of Value into question, not just theoretically but also in space and time. Blocking streets and bridges and boats and trains is a familiar nonviolent direct-action tactic, one the climate movement has employed, and it functions as low-level industrial sabotage, stopping up capitalist society's circulating flows for a relatively short while. Movements occasionally employ the strike tactic in a similar way, not to build workers' power within a workplace per se but to temporarily sabotage it by withdrawing. Swedish high school student Greta Thunberg started the school-strike-for-climate movement in 2018 when she called for students to abandon classes on Fridays, which has helped inspire a series of global climate-strike days meant to extend into the workplace. Though these tactics are clearly within the liberal tradition of civil disobedience, they frequently lead to violent confrontations between demonstrators and authorities as well as between demonstrators and vigilantes who are determined to get to work on time. Capital is not good at waiting.

Property *destruction* tactics aim to put capitalist machinery out of commission for a longer period of time or even permanently. In his popular book *How to Blow Up a Pipeline,* Andreas Malm explores the questions of violence and sabotage as they apply to the climate movement, from the relatively innocuous (using lentils to deflate SUV tires) to straight-up revolutionary people's war. "No one knows exactly how this crisis will end," he writes. "No scientist, no activist, no novelist, no modeller, or soothsayer knows it, because too many variables of human action determine the outcome. If collectives throw themselves against the switches with sufficient force, there will be no more flipping towards peak torture; the pain might be ameliorated. Within these parameters, one acts or one does not."[54] This theory of direct action has a lot of implications, and Malm draws strategic and tactical inspiration from the African National Congress and the Popular Front for the Liberation of Palestine, among other historical organizations that have targeted fossil fuel infrastructure. Though no big group has endorsed pipeline sabotage as a tactic, small groups of radicals have taken it upon themselves: In North America, both the Coastal GasLink and Dakota Access pipelines—planned to go through Wet'suwet'en and Oceti Sakowin territories, respectively—have faced repeated sabotage, including (allegedly) the 2020 derailment of a train carrying crude oil through Washington State.[55] Anonymous saboteurs announced a "treasure hunt" for Coastal GasLink, claiming to have drilled holes in or intentionally weakened various sections of the pipeline and inviting the company to start checking its work to see which claims were true.[56] Down the value chain, we can see something similar in the 2017 Correntina water war, during which around one thousand people in Bahia, Brazil, occupied a local agribusiness headquarters and destroyed gas-powered water pump infrastructure.[57] Even SUV dealership arson is on the same spectrum. This kind of sabotage probably does not qualify as public or democratic, but it's still social planning.

There remains, even among the most extreme twenty-first-century communists, a strong taboo against targeting human beings for violence or negligently exposing them to harm. In my estimation, this is a product of a combination of factors that include, in addition to widespread moral humanism, a theoretical understanding that capital's particular personnel aren't the real problem or a source of much potential leverage in the shareholder age. Kidnapping an oil executive won't change a lot except maybe get some of his subordinates promoted, the thinking goes. (Besides, he'll probably insist on talking your ear off about his company's "green" initiatives the whole time.) Communists have also tended to conclude that committing acts of violence isn't worth the risk to themselves or to their movements and organizations, either in terms of attention from the authorities or backlash from the public. However, some have at least implied that they are reconsidering the extreme tactic of assassination. "Unlike the high Cold War when politburos, parliaments, presidential cabinets and general staffs to some extent countervailed megalomania at the top, there are few safety switches between today's maximum leaders and Armageddon. Never has so much fused economic, mediatic and military power been put into so few hands," wrote Mike Davis before his death, in 2022.[58] Contemplating the threat posed by capricious individual leaders such as Donald Trump, Vladimir Putin, Viktor Orbán, and Jair Bolsonaro, Davis suggests that we "pay homage at the hero graves of Aleksandr Ilyich Ulyanov, Alexander Berkman and the incomparable Sholem Schwarzbard"—about as close as any living left-wing American intellectual is going to get to printing "Just shoot the bastards."[59]

We can see rose theory at work in the fully successful assassination of Shinzo Abe, the former prime minister of Japan, in 2022, which led the government to seek the dissolution of the influential Unification Church as a harmful cult (the assassin's mother had been exploited as a member). There are situations in which the law's status quo clearly violates people's natural right to self-defense, and they have no other choice but to either submit to gross injustice or take up arms and contest the law itself. But if we think about law in Val Napoleon's terms, this kind of revolt is about asserting a deeper lawlike principle than the state's monopoly on violence: Humans are equal and tend to defend themselves against individual and collective subordination. No one has the right to demand that any of us live at their particular mercy.

This is the plot of the early nineteenth-century novella *Michael Kohlhaas,* by the German romantic Heinrich von Kleist. Based on a real historical character from the 1500s, the book tells the story of a man treated unfairly by the royal bureaucracy and his refusal to accept it. The forced loan of two horses spirals out of control into social war as the titular character raises an army to lay waste to feudal injustice at its root, on principle. Twenty-first-century communists still read *Michael Kohlhaas,* which was written during the same postrevolutionary period as "Manifesto of the Equals," because it captures something essential about lawbreaking and lawmaking. At a time when protests against the Catholic Church birthed their own orders—at one point Martin Luther himself tries to reason with the implacable Kohlhaas—the destructive campaign was intrinsically creative: "From the Lützen Castle, of which he had taken possession, and in which he had established himself, he called upon the people to join him, and bring about a better order of things. The mandate was signed, as if by a sort of madness: 'Given at the seat of our provisional world-government,—the Castle of Lützen.'" *Michael Kohlhaas* tells of that point where society-rending anarchic violence and the first universal government bend to meet each other. If there can be no true peace without justice, then the uncompromising struggle for justice may be the most peaceful thing any of us can do.

Abdullah Öcalan, originator of rose theory, is a Kohlhaas-like character. As a founder of the insurgent Kurdistan Workers Party, Öcalan helped lead a Marxist national liberation struggle through the twentieth century's last decades. But stretched across four countries (Turkey, Syria, Iraq, and Iran), Kurdistan finished the '90s without a state, and Turkish intelligence agents captured Öcalan, imprisoning him on an island where he's been ever since. I imagine few leaders would take this as an opportunity to expand their ambitions, but Öcalan is by all accounts an uncommon man. From İmralı Island, he developed the philosophy of "democratic confederalism"—a non-state social-metabolic paradigm drawn from the experience of the state-deprived Kurdish people but designed for the entire world. "Revolutionary overthrow or the foundation of a new state does not create sustainable change," Öcalan reflects. "The state will be overcome when democratic confederalism has proved its problem-solving capacities with a view to social issues."[60] Centrally, that means, just as other peoples have concluded around the world, feminism and agroecology.

What sets the Kurdish freedom movement apart, however, is militancy. "This does not mean," Öcalan continues, "that attacks by nation-states have to be accepted. Democratic confederations will sustain self-defence forces at all times. Democratic confederations will not be limited to organize themselves within a single particular territory. They will become cross-border confederations when the societies concerned so desire."[61] Though he's still on İmralı, Öcalan's schematic found resonance in the community that inspired it, and the Kurdish freedom movement has largely adopted democratic confederalism—or Apoism, as it's called—including rose theory. As a result of the Syrian civil war, the Kurdish region of Rojava has acquired an unprecedented level of autonomy, and, led by distinct women's and gender-integrated defense forces, they've been able to hold territory under the banner of democratic confederalism, even against onslaughts by ISIS fighters.* The victorious 2014–15 battle for the city of Kobanî proved that a democratic confederalist army could stand strong and protect its territory, state or no state. In 2022, the Apoist slogan "Jin, jiyan, azadî" (Woman, life, freedom) broke containment in Kurdish Iran (Rojhilat) after Iranian security forces killed the young Kurdish woman Mahsa Amini, triggering intense national protests and international outcries. One difference between people and roses is that people can sharpen their own thorns.

* By contrast, the EZLN has mostly not relied on military victories to hold territory.

II.

ACCIDENTS HAPPEN

All three of the strategies in this book employ coercive force, whether it's to maintain the market's rules, impose labor's interests on capital, or abolish the capitalist metabolism in its entirety. Therefore I haven't held their violence against any of them, not in an ethically condemnatory way. Exalting nonviolence as an orienting principle for society is not a solution to any of our pressing problems, and at worst it serves as an excuse for an extremely violent status quo. No one has the information required to make educated utilitarian calculations about lives lost or saved under the various progressive strategies, and attempts to measure life that way rarely get beyond parochial apologias. But by seeking to transform society's values directly, communists acquire a special burden.* It's one they can't always bear.

* In Andrey Platonov's novel of the Bolshevik Revolution, *Chevengur*, the titular town's small Party leadership debates the execution of their bourgeois neighbors. A man named Chepurny rejects the idea that they won't be held responsible for the bloodshed as long as they don't orchestrate it: "What do you mean 'We won't be to blame?' We're the Revolution—we're to blame all around!" In the book, this is not an argument against the liquidation of the Chevengur bourgeoisie. Andrey Platonov, *Chevengur* (New York Review of Books Classics, 2024), 269.

The 2008 global financial crisis dragged the nation of Greece in its wake. High youth unemployment and a popular lack of faith in the state led to escalating tensions that burst open when the police shot and killed teenager Alexandros Grigoropoulos in the Athens district of Exarcheia. Intense riots roiled the city, and Exarcheia became an insurgent zone. Researcher Andreas Chatzidakis describes it this way: "Within Exarcheia, the city's traditional core of intellectual and political activity, various grassroots movements and collectives started experimenting with here-and-now politics such as solidarity trading with Zapatistas and like-minded local producers, various squats of private and public spaces (including guerrilla parks), new producer and consumer co-operatives, anti-consumerist bazaars, collective kitchens, no-ticket cinema screenings, among others. Spaces of rupture and cracks in the capitalist system began to appear everywhere."[62] Groups squatted buildings, helping themselves to a place to live. (Later, after migration to Greece increased dramatically in 2015, Exarcheia became the center of Europe's migrant solidarity movement when self-described anarchists offered full participation in communal life in the squats as an attractive alternative to the supposedly charitable refugee camps that nongovernmental organizations set up.[63]) Western leftists went east to learn from Greek comrades and brought lessons back home, providing an important ingredient for the brew that would become known as the Occupy movement. And in 2010, Greek communards pushed their nation to the edge of revolution. What happened next provides an important caution for communist politics.

After the financial crisis turned into a sovereign debt crisis, the so-called troika of international institutions (the European Commission, the European Central Bank, and the International Monetary Fund) organized a bailout for Greece in 2010, conditioned on the government's imposition of austerity measures. Here was neocolonial structural adjustment rebounding on Europe after decades of its using the same playbook against the Third World. The people knew the deal meant austerity and popular misery: Labor unions called for a general strike, and hundreds of thousands of people emptied into the streets of Athens to oppose the deal, led by the radical left. As they marched on the government seat, some insurrectionists set fire to a bank branch. No one was supposed to be at work that day because of the strike, but, fearing for their jobs, more than twenty workers were inside. Three bank workers died after suffocating on smoke, leading to deep self-criticism within what organizers had hoped would be a burgeoning revolutionary movement. Instead, a public-power strategy formed behind the new electoral party Syriza, which won power only to overrule the public and collaborate with the troika.

In the arson's wake, an anarchist collective associated with the movement—the Conspiracy of Cells of Fire (in English)—released a pamphlet attempting to reckon with the deaths. "And Death Will No Longer Have Any Power: Concerning the Fire in the Marfin Bank" is a thoroughly unsatisfactory document. The collective refuses the "American army" language of "collateral damage," but "on the other hand we will not pretend to commemorate the death of three people who, as regrettable as it was for their families, would again be a sterile news information of the system were it not the result, in the specific place and time, of a revolutionary practice."[64] The authors correctly point out that the system could never bear the same level of scrutiny, the responsibility for every accidental death produced by capitalism—never mind, I would add, the responsibility for every Alexandros Grigoropoulos—but they refuse to hold themselves and their comrades to a substantially higher standard. The struggle between communist values and capitalist Value is violent by nature, but that does not mean all consequences are justified or fall at the system's feet. If uprising is also a kind of social planning, who wants planners who not only accidentally kill workers but also aren't all that concerned about it? We have that, so to speak, in the fridge at home.

In his book *Twenty Theses on Politics,* the political philosopher Enrique Dussel tries to differentiate between "liberation praxis," which is illegal but legitimate, and "anarchist action," which is both illegal and illegitimate. He recognizes that there are situations in which death is on all sides and it becomes "necessary to discern priorities... on a higher and more concrete level of complexity, in which principles conflict with one another"—for example, "the death of the attacking enemy in a defensive patriotic struggle is justifiable... and does not stand in opposition to the material principle of life," which is of overriding importance for Dussel.[65] The consistent way he chooses to separate liberation praxis and anarchist action is "the consensual, collective, and critical support of the new system of legitimacy," or the "critical consensus of the social movement or political actor."[66] Though these phrases are alliteratively attractive, it's not clear to me that they provide a reliable guide for ethical action. Burning down an empty bank branch may reach the level of critical consensus, and accidentally killing three workers may not, yet the two are inextricably tied by the contingency of real events. Voting to excuse yourselves doesn't seem much better—maybe worse. As an impersonal system, capital can shrug off responsibility for its accidents; communists have no such recourse. If you're founding a new law, there's no one else to blame. But if laudable liberation praxis turns into condemnable anarchist violence as soon as something goes wrong, then it's always just a matter of time before a communist project succumbs.

That's what happened in Seattle, Washington, after the 2020 George Floyd uprising led radicals to occupy a section of the city known as the Capitol Hill Autonomous Zone (CHAZ), where activists expelled police and Exarcheia-like projects thrived for a month.

Living in a flash point for conflicts with the fascist right, Seattle's radical left had insisted on being relatively well armed, and security patrols roamed the CHAZ brandishing weapons. At three in the morning one night at the end of June, CHAZ security shot two people driving a stolen car who may have rammed a barricade. First reports suggested that the zone had protected itself from another fascist attack, and supporters (including me) applauded—collective self-defense, rose theory in action, a demonstration that radicals could and would protect one another by any means necessary. Then came the sickening realization that the people who'd been shot weren't Nazis on a mission but two local teenagers—ages sixteen and fourteen—and the older one was dead. The police cleared the CHAZ, but to my mind it was the failure to reckon with this shooting that doomed the zone.

These killings were accidental in that they did not represent the shared goals and intentions of the people involved. And yet they were some of the strategy's predictable outcomes, outcomes for which the movements were unprepared. A communist project that picks up a gun without a deeply rooted sense of collective responsibility and the mature capacity to earnestly grieve and learn from its mistakes shoots itself with the first bullet.

UNPOPULAR, BROKE, AND LANDLESS

If legitimacy marks the difference between revolution and terrorism, where does communist legitimacy come from? Marketcraft and public power get theirs from the traditional events called elections, whether held among the citizenry at large or within a democratic labor movement. That's a relatively uncontroversial form of legitimacy, but it is a contingent one. Voters are unpredictable and don't always act in their own interests—not the way we've been using the term *interest,* at least. Voters can (and are perhaps wont to) withdraw legitimacy from a strategy when it hits unexpected bumps in the road, even if they're already riding in the best cart. Communists, on the other hand, derive their legitimacy from their adherence to and elaboration upon invariant truths—including FETE, RCEE, rose theory, earthcare, and the equality of all people(s)—and the effective resolution of social problems thereby. In many social-historical situations, these principles have put communists on the wrong side of electoral and/or popular legitimacy. That does not rule out the strategy any more than any of the other critical sections do theirs, but it is a problem worth exploring.

The unpopularity or illegitimacy of communism is in some ways a product or even a goal of the global order Timothy Mitchell describes in *Carbon Democracy.* Starting in the period when the Conspiracy of the Equals vented their frustrations, Mitchell writes, "the advocates of representative government had seen it not as a first step towards democracy but as an oligarchic alternative to it, in which the power of government was reserved to those whose ownership of property (the control of land, but also of women, servants and slaves) gave them power over the point of passage for the revenues on which government depended, and qualified them to be concerned with public matters."[67] Mitchell points out that representative democracy still functions much the same way, with electoral legitimacy for the core and legitimacy by force for the periphery. Despite the alleged bourgeois revolution, we live in an era when kings can and do call up their jet fighters. With the world political community so split and many still caught under the authoritarian boot, it's hard to configure legitimacy for a universal program, no matter how adaptable to particular ecologies. "All people(s) are equal" is a good rejoinder to "Who voted for you to abolish the border?" but it may not be a popular one, at least not inside certain borders, which is literally what counts. It's not fair that the political legitimacy deck is stacked against communists, but that is the case.

Not only do communists ask almost all people to accept the destruction of their way of life, they also want us to actively participate in every part of that destruction. Abolishing capitalism is one thing, but now everyone has to be an agronomist? And go to long meetings? Communism asks a lot of people, especially in the areas of thoughtfulness, cooperation, patience, and self-discipline. These are not temperamental characteristics that the capitalist world has seen fit to incentivize or cultivate. Shifting from a social metabolism based on convenience to one of critical thought and action is overwhelmingly rational, given our situation, but it's not exactly an easy sell to populations accustomed to the former. Though communists seek a higher freedom, they ask people to constrain themselves in the near term. In the twenty-first century's most advanced communes, the communards have established intense requirements, such as a ban on romantic relationships for fighters in the Apoist defense forces and restrictions on alcohol consumption in Zapatista territory.* "As a social function," writes Dilar Dirik in her book *The Kurdish Women's Movement: History, Theory, Practice,* "the abolition of sexual relations in the realm of the guerrillas/cadres in favour of new forms of revolutionary intimacy based on comradeship and sacrifice disrupts, mainly in favour of women, an important realm used to mobilize and establish power and hierarchy in traditional Kurdish society."[68] Asking everyone for heavy, destabilizing sacrifices from a position of unpopularity and illegitimacy sounds so hard that it's surprising it has ever worked anywhere.

Verve is the starter motor for communist projects, and thoughtful, devoted communards are surprisingly good at working with

* "The EZLN's leaders initially wanted to prohibit alcohol to protect their clandestine organization and plans for the uprising. Women's forceful opposition to alcohol facilitated the passage of this Zapatista law, but also changed the terms of the debate." Hilary Klein, *Compañeras: Zapatista Women's Stories* (Seven Stories Press, 2015), 65.

others, even across lines of difference. Just as the TdP occupiers can't get started planting without a stock of food to eat, communists accumulate a supply of determined enthusiasm before jumping into action. They lead by example, taking the riskiest and least desired jobs within a site. It's something to behold, but verve is a difficult resource to maintain and plan around, never mind pay bills with. In the summer of 2019, a crew of Appalachian communists joined laid-off coal workers at the Cloverlick mine, in Harlan County, Kentucky, who were blocking trains full of product until they received their last paychecks, which the bankrupt company was withholding. With communist help, the protest became a protest camp. As one of the participants described it, "There's four of us that have been here consistently, and then there's a lot of other people that cycled through. We got to work pretty quickly, setting up infrastructure, a kitchen, a kids' tent, a little medical area, back-to-school supplies, just different sorts of things like that."[69] They also assumed responsibility for anchoring overnight shifts, helping miners spend more time with their families without surrendering the fight. But the strict communist insistence on universal equality ultimately divided the miners and their supporters: The communists had no choice but to make good on a threat to withdraw if a Nazi trucker who also sought to organize within the blockade camp wasn't excluded from the space. It's hard to run on enthusiasm when the people you're standing with don't have your back, too.

In terms of developed material capacity, most of today's communists have significantly less to offer the people than even austerity-ravaged states do. Venezuelan campesinos are taking initiative with Pueblo a Pueblo, but the infrastructural reality is that they're dependent on fossil fuels, and the state is the best way to facilitate access. It's not impossible to imagine an insurgent refinery commune—that's the premise of Joana Pimenta and Adirley Queirós's 2022 movie *Dry Ground Burning*, about a (fictional) gang of women in a favela outside Brasília who tap into a pipeline, pump oil to the surface, and refine it into petrol, which they sell at a discount to an affiliated group of moped delivery workers. Together, they can take on the authoritarian government. This setup is cool to watch but implausible, not least because no one should live in an oil refinery. And if campesinos in the mountains need to drive to a doctor, they probably can't wait for the communists to set up a refinery or even just a clinic with reliable supply lines. Communes are mostly not sustainable within capitalism, which means they're probably better off hijacking fuel tankers than setting up guerrilla refineries—not *Dry Ground Burning* but Vin Diesel's 2009 short film, *Los Bandoleros,* which sees the blockbuster *Fast and Furious* crew freeing a political prisoner and plotting to put their skills to work for the poor people of the Dominican Republic. "At the end of the day, people are going to get what they need," Diesel's Dom Toretto tells a populist politician who's in on the plan. "You can't move forward without fuel. And no one wants to be left behind." That's a real problem, and communists need more than movie answers.

When they're not hijacking capitalist tankers, communists often find themselves asking for sacrifices from people who have the least to sacrifice—it is not a social strategy that appeals to many people with the most, after all. While the other strategies rely on access to substantial flows of resources, whether state revenues or union dues, communists, as we've seen, worry about stocking subsistence levels of potatoes. This necessity births a lot of invention, but the mortality rate is high. It's low-stakes to proclaim "Everything for everyone!" when you don't have anything in the first place. And if your movement has to figuratively and sometimes literally fight for what it has all the time—including and especially primary mediations such as shelter, food, medicine, and freedom from incarceration—there's a limit on the remainder of care left to go around, the free abundance of which is supposed to be the commune's greatest weapon. Communes threaten to gather rather than solve capitalism's problems, drawing together people with desperate needs and few resources to assist them. That's what happened at the communized Sheraton in Minneapolis, which could not manage the collective discipline necessary to continue. As M. E. O'Brien writes of protest camps, "They require an enormous quantity of uncompensated collective volunteer hours, often from people who depend on the camp for their material survival while organizing."[70] That's a delicate balance and not something that anyone can nail down in perpetuity. O'Brien concludes that the commune can't simply spread because people next door see how well yours is working: "The structural conditions of proletarianization, market dependency, and state violence make such a vision of a revolutionary transition impossible without generalized insurrection, and leave such communities inherently unstable, inaccessible, and isolated," she writes.[71] Communism is a martial plan, and that could be unappealing to people who mistakenly believe they live in a peaceful society. Perhaps communism only seems reasonable in the mad light of unexpected fire.

LA MATANZA

The biggest threat to communist organizing is that some group of guys associated with capitalists and/or the state will disappear the organizers. The verb acquired that usage as survivors tried to make sense of what happened to their loved ones as a result of the anticommunist purges of the late twentieth century. To disappear people is to assassinate them without ceding even a dignified mourning to their communities—to leave them always wondering but forced to fear the worst. It's a tactic that dehumanizes everyone involved, a social mechanism for turning heroic people into garbage to be burned or buried in an unmarked pit. As threats to the law by nature, communists are only ever provisionally entitled to the protections the law provides for internal dissidents. Authorities use the label retroactively; in some places, the quickest way to become a communist is to get shot by the police or the army for no good reason. *Communist* comes to refer to people against whom capital and the state need not restrain themselves.

I could spend the rest of this chapter detailing the untimely deaths of people previously referenced in this book. I'd start with Gracchus Babeuf, first among the Conspiracy of the Equals, who was guillotined in 1797. The fictional Kohlhaas was beheaded, too, which is still better than what happened to the historical figure, who was tortured to death on "the wheel." Anticommunist violence was a continual feature of the nineteenth century: The monarchs put down the European uprisings of 1848 with merciless force, just as the army slaughtered the Paris Commune in 1871. In 1919, Rosa Luxemburg contemplated the European communist movement's résumé of defeat as she agitated against the war and for revolution in Germany:

> What does the entire history of socialism and of all modern revolutions show us? The first spark of class struggle in Europe, the revolt of the silk weavers in Lyon in 1831, ended with a heavy defeat; the Chartist movement in Britain ended in defeat; the uprising of the Parisian proletariat in the June days of 1848 ended with a crushing defeat; and the Paris commune ended with a terrible defeat. The whole road of socialism—so far as revolutionary struggles are concerned—is paved with nothing but thunderous defeats. Yet, at the same time, history marches inexorably, step by step, toward final victory! Where would we be today *without* those "defeats," from which we draw historical experience, understanding, power and idealism? Today, as we advance into the final battle of the proletarian class war, we stand on the foundation of those very defeats; and we can['t]

> do without *any* of them, because each one contributes to our strength and understanding.[72]

This was the last thing Luxemburg—one of the era's great leaders, perhaps the only peer of Lenin within the movement—wrote. German soldiers grabbed her, knocked her unconscious, shot her in the back of the head, and threw her body in the Landwehr Canal, where it was never found.* With the Social Democratic Party's support, the German state defeated the communist general strike and pushed the country into the first of two disastrous world wars. Innumerable comrades of Luxemburg's have died similar deaths in the century since, including as the first targets in the Shoah. The Soviets saved themselves from a similar fate by transforming the state into a war machine—first against a counterrevolutionary imperialist bloc and again against the Axis powers—sacrificing the project's communist character in addition to millions of lives. Still, it was in the name of anticommunism that European governments spent much of the twentieth century disappearing people. This Cold War era began before World War II ended, when British troops opened fire on communist-aligned demonstrators in Athens, killing more than two dozen people and inaugurating a civil war and a new totalitarian era in Greece. In the United States, where the communist movement was relatively underdeveloped, the authorities helped keep the lid on by targeting the country's most advanced communist organization, the Black Panthers, with an extralegal repression strategy, including the assassination of the deputy national chairman, Fred Hampton. Communists aren't even safe in death: After decades of attacks on his grave, Hampton's comrades bought him a bulletproof metal headstone.

* For a detailed account of the assassination of Luxemburg and her comrade Karl Liebknecht, see Klaus Gietinger, *The Murder of Rosa Luxemburg* (Verso Books, 2019).

Once again, there's absolutely no cause to constrain our analysis to the West—Luxemburg certainly didn't. Following Leanne Simpson, we can understand more than five hundred years of global colonialism as a relentless attack on communal ways of living by protocapitalists and the capitalists who emerged to follow in their murderous, thieving path. Malcom Ferdinand describes an "unbridled fire that has roamed the Earth, ravaging bodies and landscapes along the way" since 1492.[73] In the interests of analytical accuracy alone, we can't see those bodies as mere objects of victimization: Colonized peoples have struggled and continue to struggle against the imposition of extractive and exploitative regimes and have pitted their values against Value. They've fought on the battlefield and in the courts that have been imposed on them. In Luxemburg's optimistic model, this struggle has not yet succeeded. Capitalists are very good at killing people.

Though the slavery system of production officially ended in the Western Hemisphere in 1888, with abolition in Brazil, the mode of the colonial plantation continued well into the twentieth century. This model was predicated on depriving most people in the world of any say in the planning of their own societies and the imprisonment or murder of anyone who tried to make things different. In 1932, an alliance between the Communist Party of El Salvador and the Indigenous Nahua peoples launched an uprising against the country's new capitalist military dictatorship, killing dozens. The military, which had already started rounding up individual communists, responded by slaughtering tens of thousands of campesinos in what's known as La Matanza, or the Massacre. La Matanza is notable but not unique: Around the world, regimes responded to decolonial uprisings with brutal, disproportionate force, both targeted and blanket. These insurgents were increasingly aware of one another, and many had begun identifying themselves as communists in the wake of the Bolshevik Revolution.

And if they didn't, their government was likely to label them that way regardless.

It was only with the end of the Second World War—won in large part by colonial troops—and the ensuing global shakeup that they toppled the colonial mode. Victorious revolutionaries, including Mao Zedong, Fidel Castro, and Amílcar Cabral, sought to synthesize precolonial knowledge and social struggles with contemporary science in order to orient their postcolonial societies.[74] Capitalists responded with the best tools they had, and no left-wing leader could feel safe during the Cold War. Mao and Castro both evaded assassination, but Cabral wasn't as lucky. Neither was Patrice Lumumba, in the Democratic Republic of the Congo, or Salvador Allende, in Chile, or Thomas Sankara, in Burkina Faso. Coup victim Jacobo Árbenz of Guatemala died in exile, while Mohammad Mosaddegh of Iran died under house arrest. Capitalist-aligned postcolonial regimes refined Matanza-style tactics against communists and peasants, equipping their secret police forces with US-made signals technology and experimenting with terror tactics, including mass rape, torture, and child stealing. The world filled with massacres of various sizes, some known by proper names: Gwangju, Sharpeville, El Mozote, East Timor. The United States helped coordinate this effort around the world, perhaps best formalized in Operation Condor, which was an organizational clearinghouse for anticommunist repression in South America in the '70s and early '80s. As Rita Laura Segato put it about state violence in Guatemala in the '80s, it came "from the manual."[75]*

In his book *The Jakarta Method,* journalist Vincent Bevins describes Indonesia's anticommunist campaign of the mid-'60s—one of the

* Though the United States managed to contain most of this violence outside its borders, one exception was the 1990 assassination attempt on Earth First! forest protector Judi Bari, whom we met in the previous chapter: She survived after a planted pipe bomb went off in her car.

world's most deadly—and the way its revolting success helped shape a model for the postcolonial era:

> The reality is that the white world, and the countries that conquered the globe before 1945, remain very much on top, while the brown countries that were colonized remain on the bottom.... It would be too much to claim that this is because of the Cold War, or more specifically because of the loose network of anticommunist mass murder programs that the United States organized and assisted. But it's true that the period of the Cold War and its immediate aftermath, the period in which the US made routine violent interventions in global affairs, was not marked by a drop in the power of the white countries.[76]

In the twenty-first century, though capitalism declared victory in the Cold War, postcolonial massacres continue. Most of the time, it's not worth it anymore for regimes to label their Indigenous rebels as communists, but as I've explained, Native people continue to fight for communal ways of life, and capitalists continue to shoot them.

In 2016, some women intellectuals and activists gathered in Buenaventura, Colombia, for the International Forum on Feminicides. The participants were responding not to the persistence of patriarchy per se but to a global offensive against women, particularly women who stand on the front line on the other side of capital's advancing frontier. Whether at the hands of corporate militias, state militaries, or some mix of the two, women participating in communal struggle are uncommonly vulnerable to organized violence. Betty Ruth Lozano Lerma describes this as the "new witch hunt": "Many women are killed because they defend the commons, they defend the land, the forests, the rivers, and they also defend a different conception of life, in which security for the future, for beauty, consists of conserving the trees, the waters, Earth, etc. So, it is also a cultural struggle over values."[77] Silvia Federici, who has done much to popularize a communist analysis of the witch hunt, argued that "we are witnessing an escalation of violence against women—especially Afrodescendent and Indigenous women—due to the fact that 'globalization' is a political project of recolonization. It is intended to give capital uncontested control over the world's natural wealth and human labor, and this cannot be achieved without attacking women, who are directly responsible for the reproduction of their communities."[78]* Their killers do not tend to face justice; they tend to gain power.† That is a problem for the communist strategy.

If communists try to make a different kind of life at the edge of capitalist society, that makes them easy to cut out. And if communists are willing to use force to protect themselves and one another, capitalists and their states are typically *more* eager to use *more* force to repress them—not to mention the fact that they are better armed. As the strategy most likely to get you dumped in a canal or buried in a mass grave—Mike Davis estimates "at least 7–8 million dead" in the twentieth century—communism has some recruiting challenges compared to strategies that ask most people to vote every once in a while.[79] And while there is no mass of bodies heavy enough to permanently secure an intrinsically unstable social metabolism such as capitalism, system personnel have proved that killing can put off a reckoning for a long time. I don't expect them to stop: As participants concluded at Buenaventura, the authorities are speeding up their crackdown. "First

* Federici is using the term *globalization* in the conventional way rather than as I've been using it, in the sense of David Graeber's globalization movement.

† These killings have been overwhelmingly concentrated in Colombia, the Philippines, northern Brazil, Mexico, Honduras, and Guatemala. For dozens of accounts of murdered women environmental defenders, see Joan Martínez-Alier, "Women Environmental Defenders Killed Around the World," chap. 4 in *Land, Water, Air and Freedom: The Making of World Movements for Environmental Justice* (Edward Elgar Publishing, 2023), 68–90.

they came for the communists," Martin Niemöller conceded in his famous poem, which has always been true of more than just the Nazis he was writing about. That is a problem for communism, communists, and anyone who cares about them.

By itself, the communist tactic of picking a fight to the death with a stronger, more vicious enemy is what the basketball analysts on ESPN call a low-percentage shot. But communists are not by themselves: They exist in the context of everything else I've examined, the whole left field. And sometimes a fight is what everyone needs.

CHAPTER 4

THE PLANETARY CRISIS

I.

THAT IS WHAT IT IS

With a decent idea of the relevant strategies—their strengths as well as weaknesses—and the need to piece them together into some meta-strategic form, the task seems straightforward. So far I have assumed that proponents are able to enact their strategies on their own terms, addressing our social problems as subject to object. But the planetary crisis is a moving target, and it's moving a lot faster than anyone trying to catch it. Progressives are walking up the wrong escalator: No matter how many steps we take toward a particular solution, the Oil-Value-Life machine keeps dragging us down, away from our goal and toward the infernal bowels of global catastrophe. Every day, things get worse, not better. These are the unchosen conditions under which we're compelled to compile a future history of our collective choosing, and they are undeniably bad.

These days, it's easier to deny the scale or scope of the problem than to claim everything is just peachy. General peachiness levels are clearly low, but by breaking the situation into ostensibly manageable chunks, it's possible to deny the existence of a planetary crisis as such. When the situation is viewed this way, society faces a series of related but distinct and solvable problems: Decarbonizing transportation is one thing; decarbonizing energy generation is another. Decarbonizing industrial manufacturing and cleaning excess carbon from the atmosphere are tough tasks, but they're not a death sentence for humanity. Anyone who thinks the world is ending is hysterical, which means the world *isn't* ending, which means we have time to approach our problems pragmatically, one by one. The Oil-Value-Life chain is an unavoidable fact, but it's possible to introduce certain improvements consistently, with a reasonable expenditure.

These self-described moderates recall Mayor Peter Stockmann in Henrik Ibsen's 1882 play *An Enemy of the People.* Faced with a scientist—the mayor's brother, Thomas—who claims that the water for the town's signature spa baths is contaminated, Peter doesn't deny that there might be a problem. Instead, he denies the urgency. "I believe you exaggerate the matter considerably," Peter tells his brother. "A capable physician ought to know what measures to take—he ought to be capable of preventing injurious influences or of remedying them if they become obviously persistent.... I have not, as I remarked before, been able to convince myself that there is actually any imminent danger." Like the capable physician, political reformers ought to be able to remedy social injuries in accordance with the system's current general operation. That is, after all, the premise for the existence we already share. Thomas can't believe that his brother can't believe it ("It is impossible that you should not be convinced. I know I have represented the facts absolutely truthfully and fairly"), but that's not really what they're arguing about. "And what are we to do with the Baths in the meantime? Close them?" Peter asks rhetorically, understanding the situation just fine. "Indeed we should be obliged to. And do you suppose anyone would come near the place after it had got out that the water was dangerous?" By refusing the frame of catastrophe, the mayor wills the ground under his feet to stay put. "The public doesn't require any new ideas," he says. "The public is best served by the good, old established ideas it already has"—as well as, the implication goes, the leaders it already has. If the situation doesn't fit within our leaders' problem-solving capacities under the existing framework, then we must have apprehended the situation incorrectly.

"Yes but, Peter," Thomas protests, "that is what it is."

Just like Ibsen's Mayor Pete, we are all capable of believing the particular facts that confront us—about human-caused global warming, about the projections for sea-level rise and forest fires and floods—while still denying the immediacy and cohesion of the problem. Breaking up a catastrophe into bite-size pieces looks a lot like the first step toward solving it. When I was a kid, Americans talked about the environment as a question of many millions of individual choices and actions: As long as we all pick up after ourselves, the litter problem is solved. At the end of episodes of *Captain Planet* cartoons, characters broke the fourth wall and taught us children how to turn off the water while we brush our teeth and clean up the beach so the plastic loops that collar six-packs of beer and soda don't choke marine life. Many left-wing environmentalists blamed polluting corporations and warmongering governments and imperialist international organizations, but the dominant thinking held that we should divide even our biggest problems into actions so small that any American child can perform them—so small that they dissolve in the sea of society.* For the adults, who would not have the time and attention to run around turning off unnecessary lights to save energy, there was "conscious consumption"—a bastardization of Cold War labor solidarity efforts.[1] Instead of supporting liberation struggles abroad and blockade-running US foreign policy, these politicized shoppers did their best to save the world by buying recyclable packaging and hybrid cars. As I mentioned

* See Donna Lee King, *Doing Their Share to Save the Planet: Children and the Environmental Crisis* (Rutgers University Press, 1995).

above, this market fundamentalist strategy, with its dispersed nudges and consumer incentives, has failed and is only worth acknowledging as historical context.

Global warming was the inconvenient truth that dispelled the idea of environmentalism as a matter of individual responsibility. "Reduce, reuse, recycle" isn't going to halt emissions, never mind pull sublime amounts of carbon dioxide out of the atmosphere, and it is impossible to pretend otherwise. That should have spelled the end of "the environment" as an isolated and relatively unimportant policy field. It was clear to the public by Barack Obama's election in 2008 that global warming could not wait at the bottom of the national priority list—though it has somehow continued to hover in that region regardless. This unconscionable deprioritization is in large part the accomplishment of a well-funded denial-and-delay complex, a successful plan executed by oilmongers like the ones I described in the introduction. But I don't think they could have gotten away with it on their own.

Like Ibsen's mayor, we've mostly acceded to an administrative framework in which we weigh the benefits of dealing with our environmental problem against the costs and proceed stepwise according to the numbers on the scale. On its face, that sounds uncontroversial: How else is anyone supposed to solve social problems without causing a bunch more? For example, compared to his brother, Thomas is impractical. His plan to shut down the baths doesn't weigh out against the medium-term costs to the town. Peter is not *wrong* per se; he's doing his job, not so different from the ones our leaders are doing now. Just as crowds won't flock to polluted baths, voters probably won't line up to reelect a mayor who wipes his own town off the map, even if he has a really good reason to. (In Satyajit Ray's 1990 film adaptation of the play, it's a temple rather than a spa that is polluted, suggesting that pecuniary motives are not the only ones subject to threat by a novel crisis; there's no guarantee that religious and cultural values will point us toward action rather than inaction, either.) It's not hard to see that this perfectly logical conservative path leads somewhere illogical—in our case, without much exaggeration, to the end of the world. And yet it's Thomas, not Peter, who is the titular enemy of the people, much in the same way that climate protesters are enemies of the people, too, when they block traffic or interrupt public events.* Their actions don't weigh out against the near- and medium-term inconvenience of their fellow citizens, and we don't have any particular reason to believe the necessary ones ever will. Carbon democracy, as Timothy Mitchell calls it, works.

What, then, is left?

If democracy's conflicting incentives are unable to come up with what we already know is the right answer, then maybe we should start with the right answer and work backward. Some have suggested that climate authoritarianism could save the day, with a green dictatorship ordering a top-down reevaluation of values in the true common interest, whether people like it or not. But imaginary enlightened leaders (or parties) still need mechanisms to effect their visions, and those mechanisms have to operate in the real world. And the real world, as we've seen, is full of countervailing impersonal, supernational forces that undermine the ability of political leaders to act decisively on the scale required, not least because any viable solution requires the conscious participation of the world's popular masses. No person or homogeneous committee has or could have anywhere near the knowledge required to do the kind of planning we need. And just because people aren't allowed to vote on their leaders' actions does not mean they play no role in decision-making; even dictators operate with the knowledge that their authority depends on a careful balancing act between the

* Soon after I wrote this chapter, Extinction Rebellion protesters interrupted the second act of a Broadway performance of *An Enemy of the People*. The cast did not seem to appreciate the intervention, and the enemies of the people were removed.

strength and reliability of the security forces and popular will. Insofar as strong governments have led the way on experimental planning policy, it's mostly been thanks to imposed restrictions from outside—principally, economic sanctions, as in Cuba and Venezuela. There's a reason no one says that unchecked power is the mother of invention: It's not.

The climate problem is world-scale because the climate is the world's climate. And yet the world as the world is no one's responsibility. Every official and policymaker on the planet is like Ibsen's mayor: responsible only for a part, only for a time. And the best thing any given individual can do to prevent temporary local warming is, ironically, pump the air-conditioning. When it comes to addressing a planetary crisis—that is, a crisis at the level of the total ecosocial metabolism—our solutions apparatus is less than the sum of its parts, and less than adequate.

Even readers who believe everything I've written so far could be tempted—by force of habit if nothing else—to imagine they could do their jobs or follow their own strategies well on their own terms, but that would be a mistake. There is nothing that even the most "capable physicians" can do from the perspective of their jobs to contribute to solving a total problem—that's what the doctor was trying to tell the mayor. You might be able to heal a few people, or get an electric-car charging network up and running, or even blow up a pipeline, but it's only within the context of a coordinated effort *at the level of the problem* that our actions begin to address our intended object. That's why even the earnest climate action of today feels so obviously ersatz, both to everyone involved and to everyone watching from the sidelines. That's why we have politicians acting like the migration of people away from places where they can no longer live is a problem rather than part of a solution. A reasonable observer might begin to doubt that there really is a global crisis after all, since so few people seem to know how to begin acting like it.

We the people of the earth need leaders who are willing to assume a new perspective, which is to ask them to find ways to do things other than—more than—their jobs. And not our leaders alone: We ask one another to be willing to violate our partial oaths to our particular duties, to our communities, cities, states, nations, and even, perhaps, our era in order to act in the interest of the totality. That is, for lack of a more controversial phrase, an insurrectionary globalist conspiracy, albeit one that aspires to include as many people as possible. As if the task weren't hard enough already, now I'm insisting that we not only figure out a way to do what's necessary to break the Oil-Value-Life status quo but also be conscious of what we're doing while we're doing it. It would be easier if everyone could focus on their person-size part of a vast unplanned plan, the way we do under capitalism. If everyone were impersonally compelled by our social metabolism to decarbonize, perhaps we wouldn't need to plan. But alas, none of our values (including decarbonization) is isolable in a way that allows it to substitute for the thoughtful, flexible, *coordinated,* practical activity of billions.

This is a serious bummer and perhaps the biggest block standing between us and a viable strategy (or strategy of strategies) for solving our collective problems. It would be much simpler if things were different, and I worry that many of us have reflexively adopted capitalism's extreme division of labor when we think about sociopolitical change. But though capitalist boosters have stressed the efficiency of the free market as an automatic information network, the market's real efficiency comes from being able to *ignore* things the network would be better off not knowing. Is it valuable for pencil producers to know about deforestation? No: That's a waste of time at best, a fatal distraction at worst. The price of lumber transmits all the information you need. This approach has not and will not work for people committed to abolishing the metabolic order that has pushed our systems to the brink of collapse. We need to know what we're doing. It sucks, but I'm convinced it is the case.

If it is easier to persuade people to install solar panels on their houses or vote for electric vehicle subsidies or protest a fossil fuel company than to do those things as part of a worldwide revolutionary conspiracy, then so be it. We should get used to assuming difficult yet necessary tasks rather than trying to avoid them in the interest of ease. In this case, a large part of the conceptual difficulty comes from the pseudopragmatic assumption that we can and should separate the environment as a policy object from our social relations. Capitalism compels us to misunderstand our problems because the system isolates production and consumption and other metabolic moments, not just in our minds but also in our practical reality. The rest of this chapter is dedicated to suturing a handful of specific social problems to the related phenomena from which they've been erroneously sundered. My hope is that this will induce a sort of vertigo in readers as they lose sight of where the emergency stops and the stable ground of everyday life begins. *Yes,* it is difficult to believe that the whole total world is in crisis, *but that is what it is.*

II.

DANCING PAIRS

The following pages examine five "dancing pairs." The concept is inspired by a lost nineteenth-century drawing by Francisco Goya, which shows two people dancing, a man's backside in the foreground and a woman behind him, facing the viewer and her partner. Their limbs are entangled: His left arm disappears into her back, and her right juts from his torso; her left leg sneaks into the foreground as his right tucks behind, seeming to float in the air. At the edges they blend into each other. It's impossible to say just where the shadows on his pants stop and her dark dress begins. They share a border at the hip. The figures fuse, the way Plato has Aristophanes describe humans as eight-limbed monsters before Zeus split them into male and female halves (though Goya's united heterosexuals embrace front-to-front rather than back-to-back, as in the *Symposium*). We distinguish their visages, but his hat and her headscarf blend them together again. Here are two held as one: They are individual yet individual together, comprehensible in theory as independent people but knotted in practice. We can name the coupled problems that follow individually, just as we can name the dancers. We try to talk about them separately and even pursue policy solutions as if they were isolated from each other. But they are not.

CLIMATE AND MIGRATION

In 1997, the United States National Intelligence Council began publishing a report called *Global Trends*. Every four years, the NIC offers a novella-length look into its crystal ball, a big-picture account of how things in general will change in the world over the coming decades. The 2021 report, *A More Contested World,* opens with five themes, the first of which is the delicately put "global challenges":

> The effects of climate change and environmental degradation are likely to exacerbate food and water insecurity for poor countries, increase migration, precipitate new health challenges, and contribute to biodiversity losses.... Continued pressure for global migration—as of 2020 more than 270 million persons were living in a country to which they have migrated, 100 million more than in 2000—will strain both origin and destination countries to manage the flow and effects. These challenges will intersect and cascade, including in ways that are difficult to anticipate.[2]

If there's one thing you don't want your challenges to do, it's intersect and cascade in ways that are difficult to anticipate, but here we are. As we know, many countries are treating increased migration as an existential emergency. What the report projects is that, as with the climate crisis, of which increased migration is a manifestation, this is just the beginning. And you don't need a weatherman—or the National Intelligence Council—to feel the temperature rising.

At the most basic, fundamental level, global warming is directly connected to increased migration because many people are currently living in places where it will soon be too hot to continue living. In his book *The Heat Will Kill You First: Life and Death on a Scorched Planet,* Jeff Goodell describes the "widening of the thermal divide, the invisible but very real line that separates the cool from the suffering, the lucky from the damned."

> In some regions of the tropics, outdoor life will become virtually impossible. People will flee, just like many other living things, to higher, cooler climates. In many parts of the world, survival will depend not just on access to clean water, decent food, and medical care. It will increasingly depend on access to cool spaces, a job that doesn't require you to work outside during the heat of the day, and the means to escape from extreme heat events if necessary.[3]

Migration is, in the context of the thermal divide, a form of self-defense, the equivalent of a rose's thorn. As Goodell is right to note, this is not even a species-specific practice: Other living things will migrate, too, as a natural response to the changing climate.

There has been some pushback against the conflation of twenty-first-century migrants and climate refugees, but though any particular immigrant's motives are multifaceted, and though academics can concoct models to hold those decisions down and pull them apart, it is impossible to isolate the flow of populations within and across borders from drastic changes to the world environment from a historical point of view. They are a dancing pair.

"There are always a lot of reasons why people migrate," Yarsinio Palacios of the Asociación de Silvicultores de Chancol, in Guatemala, told *New Yorker* reporter Jonathan Blitzer. "Maybe a family member is sick. Maybe they are trying to make up for losses from the previous year.

But in every situation, it has something to do with climate change."[4] Blitzer describes unexpected frosts that take out whole harvests, a paucity of rain, and new pests for which farmers can't prepare. In a country like Guatemala, where more than a quarter of official employment is in agriculture, these climate changes are devastating.[5] But when the supposedly climate-focused Biden administration announced it had negotiated a "comprehensive program to manage irregular migration" with Guatemala, there was no mention of climate, just the concluding sentence "The United States and Guatemala will also deepen cooperation on border security and will continue to address the root causes of irregular migration."[6] That makes it sound like the White House is planning on solving the root causes, but if the root cause is that people can't make good lives for themselves in a particular place because of the changing climate, then migration *is* a way to address the root cause. Peeling apart the climate-migration dancing pair prevents policymakers from understanding that what was once irregular is now inescapably regular. Or, rather, it allows them to pretend.

In *A More Contested World,* the authors list the international norms they expect to endure and the ones they see teetering. Among the "norms at highest risk of weakening globally in the next decade" is "refugee nonrefoulment and resettlement"—that is, the norm that people can find asylum in new countries when they're forced to leave their homes and that they won't be sent back into danger.[7] I don't know by what metric the authors evaluate the health of such norms, but the prediction looks to be a good one. Both the Trump and Biden administrations reduced asylum protections, narrowing the right under the policy-injunctory slogan "Remain in Mexico." "In recent years," Blitzer writes, "U.S. immigration policy in Central America has largely relied on the idea that, in order to control the flow of immigrants heading north, the government should make it as painful as possible to cross the U.S.-Mexico border."[8] It has succeeded: In 2023, the UN's International Organization for Migration described this trek as the "deadliest land route for migrants worldwide on record."[9] Telling Guatemalan farmers to stay in Mexico or risk death by dehydration does not address irregular migration's root causes; it is instead a way to address its proximate cause. For some, this *is* the real climate crisis—that foreign people can get through Mexico and into the United States.*

In the European Union, states have relied in part on Greece to stem westward migration flows. To do so, Greece has employed brutal measures, including a technique called drift-backs, in the Aegean Sea. Between Greece and Turkey, the Aegean has become "not only a hotspot of state violence but also a testbed for ways to obfuscate it," as analysts with Forensic Architecture described after a review of thousands of incidents.[10] Migrants who come from or through Turkey have to navigate past the Greek coast guard, which takes that title seriously. It operates an archipelago of installations meant to prevent seaborne migrants from exercising their right to seek asylum. "Since March 2020, a new method of violent and illegal deterrence has been practiced," Forensic Architecture concluded after a review of the evidence. "Migrants and refugees crossing the Aegean Sea describe being intercepted within Greek territorial waters, or arrested after they arrive on Greek shores, beaten, stripped of their possessions, and then forcefully loaded onto life rafts with no engine and left to drift back to the Turkish coast."[11] After the then UK home secretary, Priti Patel, went on a 2021 ride-along with the Greek coast guard, England began paying the French maritime police to adopt similar "pullback" tactics in the English Channel.[12] Turning refugeedom into a sick game of Red Rover

* This does not even begin to address the emergence of *internal* climate migration, which looks to become a major social problem, particularly within the United States. See Abrahm Lustgarten, "How Climate Migration Will Reshape America," *New York Times Magazine,* September 15, 2020; Jake Bittle, *The Great Displacement: Climate Change and the Next American Migration* (Simon & Schuster, 2023).

undermines the asylum norm *as* a form of climate policy, just as the United States has at the southern border.

If we allow our policymakers to treat immigration and climate as two separate problems, it's clear they will pick the easy way out and enforce the death sentences climate change has already leveled against whole foreign populations, wielding the heat and cold and even marine currents as weapons. But using extreme weather to kill refugees is climate policy, and it's a policy we must reject by whatever means necessary.

LAND AND FOOD

Much like other living things, humans are permeable. Our environment gets inside us in uncountable ways. It flows through us—that's what it means to be part of the larger ecosocial metabolism. As both individuals and collectivities, we exist in a fragile zone between too hot and too cold, too much water and not enough. Energy from a warmed world transforms our bodies and societies, already pushing whole populations out of their homes and up against hostile borders. Temperature is only the most direct way that solar energy mingles with the human body; I turn the reader's attention to the aggregate of earth's natural cycles that we call land and its paired partner, humanity's food system.

"Food" is the concept we employ to describe the primary ecosocial metabolic mediation that we use to convert the world's energy into the edible form of calories. Traditionally, that has meant solar energy as conveyed by plants and by plants via other animals. But with the advent of fossil fuels, humans developed the capacity to excavate and eat old energy, to convert solar energy stored in oil and gas into calories in a controlled, scalable way. In the twentieth century, the Haber-Bosch process permitted the efficient synthesis of fertilizing ammonia, bringing nitrogen out of the atmosphere and into a usable form without the work of solar-powered nitrogen fixers. Fossil fuel tractors and other equipment replaced sun-grown animal muscle. In the twenty-first century, this tendency has reached a hubristic peak with indoor LED-powered warehouse farms, which seek to replace the sun itself with grid-connected light bulbs. Capital's tendency has been to divorce agriculture and husbandry from the land, culminating in the full substitution of fuel for natural context.

This is an element in the global warming crisis specifically: Burning fossil fuels for extractive agriculture destabilizes the climate, which undermines our ability to grow food in the first place. Despite the illusion of fossil fuel freedom, we're spending down a carbon budget we can't see. That framing still leaves open the possibility that we could rework our nutritional sovereignty using clean energy to run solar tractors and even LED farms. We can dream the transition from sides of fossil fuel beef to racks of electric lamb. There's no need to stop extracting our food: Instead of all that oil, we can extract minerals for solar panels and wind turbines. Clean electrification suggests the possibility of untangling food policy from the ecological crisis, and so we can deal with the two separately. But though these problems are not identical to global warming, the food system dances with the land, and the land is in crisis.

There are obvious benefits to an extractive food system—principally, that humanity can use power to bust through the limits of natural dependency. We as a species have acquired the *in*dependence to plan our lives and societies according to Value rather than our knowledge about the earth and its various cycles and how they function in particular places. In some ways, that's a much more impressive accomplishment than scooting ourselves around in dinosaur-powered metal boxes. Many people consider this kind of control over the food system to be the hallmark and foundation of modern civilization. Without it, the thinking goes, humanity would be forced to subsist at much lower population levels, scraping the dirt for whatever caloric energy the cruelly indifferent planet begrudges us. Yet extractive farming doesn't promise freedom from the earth any more than credit cards offer free money, and our bill is coming due.

In his book, *Perilous Bounty: The Looming Collapse of American Farming and How We Can Prevent It,* farmer and food-policy writer Tom Philpott describes the overwhelming ecological debt the United States has accumulated as a result of Value-based agriculture. By the grace of cheap energy in the form of transportation and refrigeration, growers have been free to chase cheap nature. As Philpott explains, that has led to the concentration of the country's agricultural output in two regions: California and the Midwestern Corn Belt. The combination of undervalued natural resources and fossil-fueled scale turned both places into specialized national organs for food production—fruits, nuts, and vegetables in the former and grains, soy, and pork in the latter. The land couldn't produce so much of the same things so frequently without synthetic fertilizers and pesticides, adding a new valence to the capitalist Oil-Value-Life chain. But fossil fuels don't actually sever agriculture's connection to the land: Rather, they offer (limited) control over the metabolic moment of *production,* turning reproduction and waste into other people's problems. "To feed us," Philpott writes, "the two regions' remarkably productive farmers are drawing down what are essentially fossil resources, at rates much faster than they can be replenished, and in a way that makes them increasingly vulnerable to weather shocks."[13] In both the Corn Belt and California's "Salad Bowl," the limits are arriving, as the Salad Bowl runs out of water and the Corn Belt runs out of good soil.

The cheap-nature cupboards are bare-bound in both places, but farmers can't stop. (Or, rather, farming capital can't stop: Individual farmers have been forced out of the industry at a rapid clip, their land swallowed up into progressively larger agglomerations.) The financiers who own California's farmland are going to juice every drop out of the state's aquifers and gulp down as much of the Colorado River as they can get their lips around. And the federal government is committed to hoisting up the already unprofitable Corn Belt to the benefit of input sellers (fertilizer, farm equipment, pesticides, seeds) and output buyers (cooking oil, pork, and, yes, oil companies, which use the subsidized ethanol as an additive). Growing international demand for Iowa pork and California almonds is in part a demand for Iowa topsoil and California groundwater—increasingly precious resources that the market knows not to measure too well. Trying to escape natural limits has merely hidden their approach. And though many firms seek to employ capital profitably in what investors might call the food space, society is not planning for the imminent collapse of the capitalist food system. That's the course of action any reasonable group of people would take under the circumstances, but the current mode of social organization makes that practically impossible.

The United States is hardly an exception when it comes to exhausting food. A global market means producers everywhere have to compete with the biggest, baddest, most fossil-fueled monocrop exporters. And if you can't beat them, you have to join them—if you're lucky enough to get the invitation. In Yemen, for example, domestic production of the traditional grains barley and sorghum succumbed to imported corn, wheat, and rice at the end of the twentieth century. Fish and seafood production has grown significantly, exploding in the twenty-first century, but not to feed Yemenis. As one scholarly account of the nation's agricultural transformation concluded, "Fish is currently not a main part of the population's diet, in spite of Yemen's long coastline and abundant marine life."[14] The country has been compelled to deplete its stocks for the world market; the fish are too valuable for the people to eat.

Analogous pressures have compelled big agribusiness to mine the water of Brazil's Cerrado savanna, turning grasslands into soybean fields for animal feed and the export market. Combined with declining rainfall resulting from climate change, this has meant drought for small and subsistence farmers of traditional crops.[15] Value's pokes and prods send Brazilians hunting for the country's cheapest nature: The Amazon rainforest is, by capitalist standards, underdeveloped. Artisanal gold mines are an increasing cause of deforestation in the region, with desperate semiproletarians harder to regulate than giant multinationals.[16] And though, in a rare act of ecological planning, clear-cutting deforestation has been reduced, smaller-scale selective logging has left forest areas degraded, with similarly disastrous ecological effects.[17] An international journalistic investigation of Brazilian beef supply chains showed just how inseparable the world's Value-led extractive food system is from environmental policy: Cattle ranchers breach the protected forest, clearing it for pasture; there, they harvest that cheap (stolen) land in the form of beef, which they sell to a supplier who sells it to an exporter who sells it into the increasing global demand for meat.[18] We can't separate extractive agriculture from the land because the land is what's extracted.

Yemenis can't afford to eat fish from the Arabian Sea, and Brazilians can't afford to eat manioc from the Cerrado. But an increasing number of people around the world can afford to eat the Amazon in the form of a hamburger and California's groundwater in the form of a pistachio. Electrification doesn't change this dynamic, which emerges from Value's domination over Life. But even if the global capitalist food system won't stop dancing with deforestation and drought, is that the same as the climate crisis? On some levels, the answer is simply yes. The paired land-food system is both vulnerable to and a significant cause of global warming. Extractive agricultural zones are, by definition, wounds to the earth, nearly certain to contribute to the destabilization of planetary cycles that we describe as climate change. Forests and savannas absorb carbon, and destroying them is akin to burning fossil fuels, even without the literal burning of fossil fuels that powers extractive farming—not to mention the methane from all those cows (and all that rice). Everything I've described is related in one way or another to the climate.

At the same time, global warming and climate change are insufficient containers for the phenomena we're exploring. These problems are at least as much social as ecological. The same song animates each of the pairs; they follow parallel steps, though they share no instruction manual. Total crisis is the dance, and somehow every beast and bird knows the tune.

ANIMALS AND WORKERS

One unseasonably hot day in the winter of 2022, I stepped through my door in Washington, DC, to the sound of birds screaming. At least that's how I interpreted the racket in the trees. And I didn't blame them. *This weather must be driving the animals absolutely crazy,* I thought. *They can feel in their bodies that things in general are going wrong, and there's nothing they can do about it.* It took me a moment to remember that I am an animal, too, that I could feel in my body that things in general were going wrong, that the weather was driving me crazy, and that every conversation I'd have that day would start with the same avian squawk: "Can you believe how hot it is?" Even if we'd prefer to think of ourselves as the inside to animals' outside, humans and nonhumans are stuck together in this planetary crisis.*

In a televisual trope sometimes referred to as the "foreboding

* "Embodied cognition implies that the mind is subject to the whims of a wild planet. You mirror your environment, and not in an esoteric Age of Aquarius sense. . . . As the environment changes, you should expect to change too. It is the job of your brain to model the world as it is. And the world is mutating." Clayton Page Aldern, *The Weight of Nature: How a Changing Climate Changes Our Brains* (Dutton, 2024), 9.

fleeing flock," human characters get tipped off to an imminent problem when they observe animals behaving strangely. Maybe the birds are flying the wrong way, or the whales are beaching themselves. We rely on other animals as an interpretive bridge to the ostensibly inanimate environment. Or at least we should. Humanity didn't heed the warning in a study published in the January 2015 issue of *Proceedings of the National Academy of Sciences* titled "Recent Shifts in the Occurrence, Cause, and Magnitude of Animal Mass Mortality Events." The researchers concluded that the magnitude of these MMEs has been "undergoing taxon-specific shifts" and that "MMEs associated with multiple stressors and disease, which are associated with the largest MME magnitudes, are increasing in frequency"—findings they dryly describe as "surprising." Though perhaps this kind of normative language is outside the bounds of objective science, a surprising increase in wild-animal mass mortality events is not good.

Only around a quarter of the MMEs in question could be directly tied to climate—the portion seems certain to rise rapidly—but we know that changes to land use, for example, are only a degree or two of conceptual separation away. It's a good way to illustrate the total nature of our emergency. As journalist Marion Renault put it, "These events ask us to consider the difference, if any, between what is overwhelming to an ecosystem and what is overwhelming to us. And they force us to confront how little we know about the world in crisis around us."[19] However, the framework such research employs poses humans as potential spoilers or stewards rather than another kind of animal that's subject to whatever the hell is going so badly. Researchers describe "human perturbation" as one of the leading causes of MMEs, but then how do we think about our own taxon-specific MME, COVID-19? As of this writing, seven million humans have died of a pandemic disease that, best as we can tell, emerged as a result of humans perturbing wild animals. As the drive for unexhausted cheap nature steers more workers into the forest, increasing crashes along the wildland-urban interface (WUI) are a serious threat to both sides.

However, if we understand this as a collision between people and animals or between society and nature, we make a couple of important mistakes. Not only do we err by dehumanizing the Indigenous communities that operate in and among the so-called wilderness, we also forget that settlement at the WUI isn't driven by people as such so much as by Value's inhuman thirst. In Sophie Chao's ethnography *In the Shadow of the Palms,* she studies the transformation of Indigenous Marind communities in the Indonesian-occupied province of West Papua as the oil palm industry has converted forestland into monocrop plantations. Palm oil expansion is one of the most egregious agribusiness causes of global deforestation—the ingredient has proved very useful in the production of ultraprocessed foods, on which an increasing number of the world's people have been compelled to rely—and the advancing frontier follows a logic that's familiar from previous pages. Marind people, Chao finds, seem to blame the oil palm plant itself for the destruction of their lifeworld, at least more than they blame the Javanese migrant workers who cultivate it. Insofar as the palm thrives while the workers get exhausted, this is a reasonable interpretation. Here the encroachment of capitalist social metabolism manifests itself as an attack on people by a plant.

In the crisis dance, nonhuman animals find themselves under the same mute compulsion as humans. As palm expansion destroys their traditional homes, West Papuan forest animals such as deer, possums, and tree kangaroos are forced into Marind villages, where they have no social role to sustain them. Chao tells the story of a cassowary chick named Ruben, the only survivor of a nest found in an irrigation ditch in a burned-out forest that was on its way to becoming a giant palm plantation. Though villagers saved the bird's life, they had no particular use for it. Rural Marind don't really keep domestic pets, and unlike chickens, the cassowary didn't serve any functional purpose. This forced the bird to scrounge on the proverbial streets for scraps of processed human foods such as instant noodles and cookies. Because he was no longer able to reproduce himself with the forest the way his parents did and was effectively unemployable in the village, we know what Ruben's position is called: semiproletarian. It does not escape Marind notice that they find themselves similarly situated. "Much as village animals are reduced to begging for food and water from their keepers, rural Marind are increasingly dependent on financial aid and food handouts from the government and from agribusiness companies to whom they have ceded their customary lands," Chao writes. "And with state endorsed agribusiness projects multiplying relentlessly, both humans and animals face the dire prospect of a rapidly disappearing forest to live in and from."[20] As one elder told Chao, "Ruben, me, Marind people—we are all the same. We are all victims of the government and of oil palm."[21] Value reduces humans and other animals to the same universal substance.

Pairing human workers and other animals gives us a different, better perspective on the opposition between capitalist Value and living things than we get from imagining the planetary crisis in anthropocentric terms, as something humans are doing or something that is happening to us alone. Consider, for example, the attack on the fish population. Worldwide, there has been an intensification of seafood production—open-capture fishing has industrialized just to keep output at a consistent level, and input-heavy aquaculture is booming.[22] Mass aquaculture left traditional agroecological techniques behind and embraced the model of industrial agriculture, especially the part about spending down the environmental nitrogen budget.[23] Meanwhile, ocean extraction is a great way for powerful countries to take advantage of smaller, poorer nations with good coastlines, whether Chinese ships off North Korea and Madagascar, EU and Turkish trawlers in West

Africa, or the US fleet making good use of its Pacific colonies. Like the crisis, industrial fishing is a global phenomenon: To feed the engorging American demand for calamari, Chinese ships impress workers from Indonesia to go harvest squid near the Galápagos Islands.[24] But though this industrial transformation has caused serious damage to marine ecosystems as per-capita seafood consumption increases, this is not a case of people versus fish. It's Value contra Life.

Humans are victims of these extractive practices, too—traditional and artisanal fishers and the communities that rely upon them *as well as* fishworkers who labor on megascale industrial boats, sometimes in literal slavery.* A 2015 Associated Press investigation into coerced fishing labor culminated in the release of more than two thousand enslaved workers who had been trafficked into Indonesia from Myanmar, Thailand, Cambodia, and Laos.[25] Instead of expecting fishworkers to live the high life at the cost of cheap nature, recognizing the fish as our underwater dance partners draws our attention to the link in our fates. Overfishing goes with overworking, and life quality deteriorates for sea creatures and hairless apes at the same time because the capitalist class finds itself encouraged to generate savings in both places, to exhaust nature and workers. Insofar as we represent costs, it is in capital's indifferent interest to reduce and degrade working people and other animals alike.

The foreboding fleeing flock is not the only way we can use animal observation to understand our own situation. If we look closely, we can discern two distinct but interrelated tendencies: There's global

---

* Though these two groups are distinct conceptually, they sometimes consist of the same people. As an Associated Press investigation into underpaid foreign fishworkers working off American ports found, "The fishermen are not just cheap, they're skilled. Many from traditional Indonesian fishing villages, like Pemalang on the central coast of Java, start going to sea as young boys with their fathers." Martha Mendoza and Margie Mason, "Hawaiian Seafood Caught by Foreign Crews Confined on Boats," Associated Press, September 8, 2016.

industrialization, in which complex large-scale production processes stretch like a knit web over the whole world, but there's also *de-development,* in which Value's calculations dictate that, in certain places, it's better to steal and destroy than to grow and build. Political instability in Somalia, for example, has opened up its coastal waters to illegal, unreported, and unregulated fishing by foreign-flagged vessels. This type of fishing has increased by more than 2,000 percent since 1981: The fastest increase followed the country's civil war, in 1991.[26] An analysis published in the journal *Frontiers in Marine Science* showed that five of the fifteen species groups pursued by both foreign and domestic operations are being fished unsustainably. Those five species make up most of the domestic catch, and overfishing them renders hundreds of thousands of lives unlivable.

Foreign fleets benefit not only from a lack of local licensing and law enforcement capacity but also from a lack of homegrown competition. "Maritime and fisheries education ceased altogether after the collapse of Somalia's government in 1991 and it has been reintroduced only recently, without proper planning, government support or regulatory oversight," writes consultant Abdirahman J. Kulmiye in his detailed report on the industry.[27] The country has been robbed of its fish, yes, but it's also been robbed of the human capital it could have developed with a share of its own resources. And, as Kulmiye points out, Somalia has never had its own industrial fishing fleet and can no longer depend on the resources of its past international developmental partners, the Soviet Union and Iraq. Now, according to Global Fishing Watch, nearly two hundred Iranian-flagged ships have opportunistically targeted underregulated waters off Somalia and Yemen in what the organization called "likely one of the largest illegal fishing operations occurring in the world."[28] Capital in the twenty-first century would rather invest in a limited-time-only up-for-grabs seafood buffet than a sustainable sector in Somalia.

At this point, a perceptive reader might imagine that those Iranian-flagged ships are not just scooping up fish for the Iranian people. After all, someone else might pay more. Oil-Value-Fish sounds goofy, but not to more than four hundred thousand Somalis for whom fish are life.[29]

Oil-Value-Sheep sounds goofier still, but it's an even more destructive chain. In a de-developmental process that researchers Mark Duffield and Nicholas Stockton call "militarized livestock production," increasing demand for sheep has led to a "deadly destruction–consumption embrace" between the Persian Gulf and the Horn of Africa—specifically, Somalia and Sudan:

> Separated by a decade or so, the social civil war in Somalia and the protracted genocide in Darfur marked the consolidation of a new militarised mechanism for acquiring, herding and marketing livestock, especially sheep. Initially conducted through camel- and horse-mounted militias from the respective marginalized norths of both countries, this violence quickly converted profits from selling on animals stolen from politically disqualified farmers, into fleets of motorbikes and Toyota "technicals", that is, pick-ups mounted with heavy machine guns, to facilitate rapid and far-reaching land clearances.[30]

Bullets become muscle becomes more bullets. But because this militarized sheep stealing is an extractive, exhausting mode of production, the researchers track the Persian Gulf's food frontier as it has moved, destabilizingly, through Somalia and then Sudan. De-development there powers environmentally absurd megaprojects across the sea; those gleaming towers don't build themselves, and construction workers in Dubai need more protein than is locally available.*

* A similar circuit pulls gold out of artisanal mines in the Sahel—Mauritania and Mali in particular—and smuggles it to Dubai for refining via so-called gold flights, leading to

De-development doesn't just feed development on the consumption side. Working people from de-developed areas play a disproportionately important role on the production side of global industrialization as well. In the past few decades, the American meat and poultry industry has consciously pursued refugees from around the world to staff its plants—people displaced from conflicts in the Horn of Africa (Somalia, Eritrea, Ethiopia, Sudan), Myanmar, the Democratic Republic of the Congo, Central America (El Salvador, Guatemala, Nicaragua, Honduras); from the War on Terror (Afghanistan and Iraq); from American sanctions (Cuba and Venezuela); and, most of all, from Mexican rural de-development in the wake of NAFTA. These are unpleasant, extremely demanding jobs that do not pay well except compared to poverty. Without an English language requirement, though, managers can scout the world for its most vulnerable laborers at any particular moment. And if your Somali workers demand prayer breaks, for example, there's always another civil war somewhere else.[31] Global industrialization doesn't mean that every region sells its local animals on the world market. Rather, Value flows push and shove people to the places where they're best exploited and modulate nonhuman animal populations according to a similar logic. A Somali sheep goes to Saudi Arabia; a Somali worker goes to Kansas to cut up cows. As one of those workers, a young man named Ahmed, told a reporter—with the authority that comes from experience—about the smell of the National Beef factory where he worked: "It's like goats in a refugee camp."[32]

Global industrial animal production reaches apogetic heights in the American Midwest, where vertically integrated pork manufacturers intertwine their workforce and their pigs in each other's biology to a staggering degree. For his study *Porkopolis,* anthropologist Alex

analogous labor superexploitation, political instability, and environmental destruction. See James Pogue, "Gold Fever in the Coup Belt: The Mines of Mauritania," *Granta,* July 18, 2024.

Blanchette embedded in an Iowa pork town that births, raises, and kills millions of animals every year. What he found was not a "project of human mastery over hogs" but rather an unfolding model of modern meat that "revolves around remaking the lives and labor of human beings to make them amenable to capitalist animality." The meat and poultry industry is difficult to mechanize and automate because, no matter how hard corporations try, it's impossible to standardize living things. Even the largest-scale firms depend on a lot of workers to make quick and precise judgment calls about something intrinsically variable, such as where to cut a subtly individual carcass or how to sexually stimulate a particular sow for artificial insemination. Instead of humanizing the industry, this dependence has incentivized firms to industrialize people. Blanchette writes of a firm that puts potential employees through tests at the company health clinic to determine where their bodies are strongest so as to match them with the part of the pig they are best suited to repetitively disassemble. When he asked workers how the disassembly line was transforming their physiologies, "some men enthusiastically lifted up their T-shirts to reveal subtly different patterns of muscular development etched across their backs, while others showed how one of their wrists, arms, or fingers differed from others in size or shape."[33] Slaughterhouse injuries have changed, from acute maimings to the more affordable chronic muscle and nerve damage. Capital intends to get the most out of its pigs and its people, no matter if what results is gruesome and obscene.

Among the gruesome and obscene transformations to the industrial pig is a phenomenon with the evocative name "systematic runting." In the attempt to cut costs, the industry has tried to maximize a metric called P/S/Y, or piglets per sow per year. Between 1989 and 2014, the average US litter size increased by more than 1 percent annually, from 7.86 to 10.31, with the top farms averaging over fifteen. The piglets, however, have been shrinking, and what's worse, sows usually only have at most fourteen working teats. These farms are intentionally producing runts—piglets that can't survive without active intervention, including bottle-feeding—thereby replacing porcine maternal attention with human labor at the level of reproductive biology. "Systematic runting makes sense only once conditions of migration, hiring, and valuation lead to abundant and cheap labor," Blanchette writes. "Unlike modernist scenes of a dominated nature, the runting of life suggests a matter of exerting incredible amounts of social knowledge and capital—in the forms of breeding and genetics—to make hogs and humans codependent at a fine physiological level, embedding work into the development of even things such as musculoskeletal structure."[34] The Value system has written social crisis into pig genes, creating industrial animals that are dependent on perpetual civil war between humans in order for their babies to survive. The systematic runting of life indeed.

As in a previous era of climate change–induced total world crisis, people and other animals are in the same boat. But instead of imagining how we can improve the ecological situation for everyone, humans insist on using human metrics to assess how things are going on the planet in general. Economics attempts to separate human welfare from the environment that provides its foundation, but it does not succeed.

NATURAL DISASTERS AND PROFITS

How should we think about the relationship between capital accumulation and natural disasters? In her book *The Shock Doctrine,* Naomi Klein laid out what has become perhaps the dominant framework on the left. For capital and its political allies, she writes, natural disasters are a great excuse to implement economic "shock therapy" and its three trademark demands: "privatization, government deregulation and deep cuts to social spending."[35] As a suite, these policies render the working class increasingly dependent on capital while simultaneously freeing capitalists from democratically imposed competition and restrictions. The shock doctrine is an attack on marketcraft and public power at the same time, and concomitant investment in police and other security forces keeps a lid on the communists. It's an attack on the whole progressive spectrum on Value's behalf. Corporate profits thrust up; environmental protections evaporate. Capital accumulation entwines with natural disaster, forming a particularly sinister fourth dancing pair.

Destruction creates desperation, and one family's desperation is some capitalist's opportunity. Investors know how to make lemonade out of a flood, or a fire, or a hurricane: In the wake of natural disasters, property sells at a discount, notwithstanding the damage. That means that buyers can make good money without developing the underlying asset just by purchasing it from immediately crisis-stricken people and holding it a little while, until the literal storm blows over. Reporting on these "vultures" for *Business Insider* in the wake of the 2023 Hawaii fires, Anthony DiMauro writes that "for disaster investors, these increasingly common tragedies are an opportunity to win big, and the residents who have lost their homes pay the price."[36] Similarly, unpredictable weather has pushed small farmers out of the business and their land into the hands of multimillion-dollar industrial operations and institutional investors.[37] For a working person, natural disasters and their vulture profiteers are a one-two punch, the second blow disguised as half a favor.

But if natural disaster and profit are dance partners, they're like the con man Gaston and the pickpocket Lily in the 1932 screwball comedy *Trouble in Paradise*. When the film starts, the two are both disguised as wealthy aristocrats, and each plans to rob the other. When they discover the truth, they start revealing what they've taken from each other: He takes his wallet back and shows her a jeweled brooch. She points out his missing watch, and he pulls her garter from his pocket. "I have a confession to make to you," Lily says. "Baron, you are a crook." It takes one to know one, and they know each other quickly.

We already know how capitalists rob nature. As the ecological theorist Vandana Shiva writes, "The market-oriented development process risks destroying nature's economy through the overexploitation of resources and the uncomprehended destruction of ecological processes.... The short-term positive contribution of economic growth from such development may prove totally inadequate to balance the invisible or delayed damage."[38] Robbing nature saves capitalists from having to get real jobs, but nature robs back. Wildfires don't always know to leave rich people alone, as the Woolsey Fire proved in 2018 by destroying $6 billion worth of Southern California beach real estate—not the right crowd to be shock-doctrined. In fact, as coastal real estate has become incredibly valuable, the worldwide natural disaster price tag has grown to more than $300 billion annually, which is in the same neighborhood as the GDP of Chile or Pakistan.[39] In 2022, hurricane damage wiped out an estimated 7.5 to 10 percent of Florida's state GDP.[40] And in 2023, major insurers State Farm and Allstate announced they would pause on issuing new policies for California homes because the climate-enhanced fire risks made the market unacceptably difficult.[41] That's too much disaster and not enough capitalism.

The cost of nature's counterattack is deeper than fire and flood damage; it goes all the way to capital's very ability to reinvest and accumulate. Because all aspects of production are anchored in the physical biosphere and its cycles, the bigger capital gets, the greater its dependence on the same systems it undermines for profit. A recent quantitative analysis published in *Nature* estimates that, because of climate change, "the world economy is committed to an income reduction of 19 percent within the next 26 years independent of future emission choices," amounting to tens of trillions of dollars a year in annual damages.[42] Creative destruction isn't necessarily a problem as far as Value is concerned, but destruction that precludes reinvestment is a genuine loss. Paying a guy to strip the copper pipes out of a particular building might be a good investment once, but not the second time. And now the water doesn't work.

Think about agriculture. We might imagine that, given the near- and medium-term risks the planetary crisis poses for the earth's food system, capitalists would be investing huge amounts in figuring out how to do more with less. And yet when researchers from the US Department of Agriculture's Economic Research Service looked at the global agricultural-output growth rate over time, they saw a huge drop-off in the 2010s. Though the decade witnessed a serious increase in the expansion of agricultural land—more than doubling its rate—this expansion didn't make up for the deceleration in efficiency, which dragged the global output growth level below 2 percent. "Overall," the researchers write, "a slowdown in global productivity growth suggests that farmers will need to use more land and apply other inputs more intensively to maintain growth in output."[43] What do they suspect is causing this slowdown? "[C]limate change and associated weather shocks such as drought have slowed the growth in or decreased agricultural yields. If weather shocks become more extreme or frequent over time [and we know they will], the negative effects on agricultural productivity are likely to become even larger." As they point out, some countries in South America and Africa have seen their agricultural efficiency growth go *negative* between 1991 and 2020, which the researchers blame on farmers expanding onto less productive land, the exhaustion of natural resources, and climate change effects—all three of which are bound to increase worldwide. But even in China, where agricultural productivity growth has been high, climate disaster poses a threat to reinvestment. A study published in the journal *Frontiers in Ecology and Evolution* found that exposure to climate shocks has a "significant impact" on levels of productive investment among Chinese farmers.[44] Rather than invest to meet the challenge of a warming world, capital is mostly holding on to itself for dear life. But scared money doesn't make money, and capital thereby starts to rot.

If capitalists rob nature, and if nature undermines capital, everyone ends up broke. Where is the money supposed to come from? It's a real concern: The World Bank called the early 2020s the "slowest half-decade of GDP growth in 30 years," driven by timid capital investment.[45] There are fanciful solutions—asteroid mining, for example, or magic software—but the likely answer is much more prosaic: us. If climate change means the future is progressively less guaranteed, then capital looks to protect itself by reducing costs instead of riskily planning to increase output or venture into new fields. For firms, that could mean consolidating with competitors, hiring consultants to "trim the fat," wheedling subsidies out of the government, increasing prices, reducing product quality and/or quantity without changing the price, or getting customers to do some of the work for them.* But the tried-and-true way to reduce costs is to increase the rate of labor exploitation. That can be as simple as neglecting to award raises to employees fast enough to keep up with the cost of living or as dastardly as union busting or wage theft or child labor. In 2021, UNICEF and the ILO issued a joint statement warning that they had measured the first increase in decades in the number of child laborers around the world. In the United States, the Economic Policy Institute tracked ten state-level bills meant to weaken child labor protections, concluding that such efforts "are part of a coordinated campaign backed by industry groups intent on eventually diluting federal standards that cover

* Researchers modeling the impact of temperature increase on global inflation found "higher temperatures increase food and headline inflation persistently over 12 months in both higher- and lower-income countries." Maximilian Kotz et al., "Global Warming and Heat Extremes to Enhance Inflationary Pressures," *Communications Earth & Environment,* March 21, 2024.

the whole country."[46] The ILO estimates that ten million additional people fell into what it calls modern slavery between 2016 and 2021.[47] By attacking the working class, capital can find a way out of its profit squeeze, and indeed we've seen workers and consumers rather than capitalists pay for the slowdown in growth.

Offloading costs and risks to labor and the public at large is a more viable solution for the capitalist class than a race to the bottom of the extraction barrel that would pit capitalist against capitalist and accelerate the earth's chaotic exhaustion. Private green investments will help reduce climate change in the future, but they're not immune to its effects now. Clean energy infrastructure doesn't get an extreme-weather exemption—consider the unprecedented March 2024 hailstorm that smashed up a giant solar farm in Texas, leaving neighbors to worry about chemicals penetrating the water table, for example.[48] Green stuff, then, is not an answer for investment capital, certainly not when it can't lay the risks off on the public. Instead, profitable innovation means finding sustainable ways to steal from people, like generative AI software that chews up genuine creative work and spits it out again in flavorless chunks, for a fee. Constant downward pressure on the working class eventually causes people to start buckling. The wage—minus the unavoidable fees, fines, scams, and other connivances—ceases to be sufficient to reproduce people's lives, forcing them to take extra jobs, to commit crimes themselves, to forage for food, to migrate. The price of labor drops below its cost, and capital increasingly finds its gains at the working class's direct expense, digging into bodies like a long rope on the bare palms of someone sliding down.

TRANSITION AND NATIONALISM

If I've done anything close to my job, it's clear by this point that the social struggle that will fill the historical epoch we hope to call the fossil fuel transition assumes the character of a class conflict by other means. Attempts to frame the situation as a fight between nations or groups of nations, between people or races, between humanity and nature, and even between everyone and the mistakes of the past all misread the current coordinates. Circumstances will force either the exploiting or exploited class to pay a decisive price in the coming decades. To capital, genuine answers (such as globally planned migration, nationalized utilities, a decline in the profit rate, and fossil fuel asset stranding) appear as problems, and disasters (such as lowering our expectations for how much warming we can avoid and intentionally spewing sulfur into the air) suggest themselves as solutions. Capitalists are incapable of implementing the limits to extraction that the workers of the world must then impose on behalf of the species in general. No matter which class declares victory, this conflict resolves in a revolutionary reconstitution of global society at large. Either that or the common ruin of the contending classes and the earth itself.

We are unable to even start truly reckoning with the disaster at the ecosocial metabolic level until one side prevails; the stakes are unavoidably high. Where will this struggle play out? Not in terms of geography—we know the answer is "everywhere, and some places in particular"—but institutionally. Within the liberal tradition, the political sphere is supposed to be where groups go to decide their contests on a democratic and nonviolent basis. Since the French Revolution, the political domain hews to a kind of neutrality when it comes to class conflict, if not in practice, then at least in theory. "Carbon democracy" pulls the mask off the liberal framework to reveal a pulsing network of oil extraction, distribution, and combustion underneath, but the apparent solution to undemocratically distributed global democracy is to, well, democratize it—to create new cooperative, supranational bodies where everyone can deal with our shared challenges on that basis. That's what the Intergovernmental Panel on Climate Change

(IPCC) is ostensibly about: getting the whole world on the same page. But so far the international effort to deal with the planetary crisis at the appropriate level has been a series of miserable failures. Instead, we've seen a reversion to what journalist Christian Parenti has called the "politics of the armed lifeboat": "responding to climate change by arming, excluding, forgetting, repressing, policing, and killing."[49] The transition thinks she's dancing with the liberal global order, but when she pulls back from its swaying embrace, she finds herself under the cold eyes of exclusive nationalism.

The kind of nationalism I'm writing about reflects capital's interests not openly but by disguising the struggle's class nature and focusing popular attention on other enemies, both foreign and domestic. The transition itself is inextricable from these political attempts to depoliticize it. Migration is a good example: Rather than deal with the needs of the global working class, rich nations have made their borders more likely to kill people. And nationalism doesn't stop at the water's edge: Human Rights Watch (HRW) has described the European Union's lending and aid facilities as an "ATM for autocrats." Regarding a €7.4 billion deal with Egypt, HRW's Andrew Stroehlein wrote, "It's very simple: you promise to stop the migrants, you get the cash. How are you supposed to stop them? EU leaders don't care! Do whatever you want! Are you worried about so-called 'European values' of democracy and human rights? Well, don't be!"[50] The result is that workers around the world are experiencing and will experience the fossil fuel transition as the imposition of authoritarianism, even if the money is coming from people who look in the mirror and call themselves liberals every night. The connection between international border funding and authoritarian politics is not vague or metaphorical; it's direct. In February of 2024, for example, Al Jazeera reported that a special unit of the Senegalese security forces funded by the European Union's Emergency Trust Fund for Africa—a pile of money "dedicated to addressing the root causes of migration in Africa"—had been used to repress opposition protests within the country, leading to dozens of deaths. "In one video, security personnel in the same type of armoured vehicles the EU bought," the report reads, "are seen firing tear gas at a protest caravan."[51] There's no such thing as a single-purpose armored vehicle.

What, for capital, is the point of killing people anywhere in the world when capitalists could exploit them in a job instead? As I've pointed out, there isn't enough capital (or good jobs) to go around. People who fall outside the system's need for laborers appear as a surplus population, more useful outside formal employment than in it. The theorist Ali Kadri describes two ways that intentionally wasting working people's lives can be productive from the capitalist perspective, both economically and politically: "First," he writes, "it deracinates, socialises, and calibrates the growth of the reserve army of labour to the rate of accumulation and, secondly, it ensures the stability of capital's rule as it disempowers labor."[52] The American meat-processing industry, where the world's most desperate proletarians risk life and limb in order to fill US slaughterhouse boots, has given us plenty of examples of the first. The second is deceptively simple: Since capital is in a direct conflict with labor, one way for capitalists to secure their place in the world is to kill members of the working class. Not all of them, obviously, but enough for everyone to get the idea. "The clout of labour in the class struggle determines the share of labour from the social product," Kadri explains.[53] The economy is political, and increasing the rate of exploitation to compensate for natural exhaustion and climate uncertainty requires a reduction in the willingness and ability of working people to fight for more. Workers need to be beaten, which means workers need to be beaten. This is itself an investment opportunity.

In the 1997 movie *Grosse Pointe Blank,* a hit man (John Cusack) goes home for his high school reunion, and his ex-girlfriend's father wants to know what he's been up to, assuming he followed his early

Gen-X slacker inclinations. When Cusack responds honestly ("professional killer"), he gets encouragement: "Good for you," the dad says. "It's a growth industry." Though it's less famous than the "Plastics" line from *The Graduate,* this turned out to be at least as good as far as career advice goes. Major defense stocks have performed extraordinarily well in the twenty-first century, catapulting over the also vibrant S&P 500.[54] And minor defense companies have had good luck being acquired by those big ones. "It's just, sadly, a good business to be in right now, given the state of affairs in the world," defense analyst Roman Schweizer told *Defense News* about the American rocket weapons industry at the end of 2023.[55] "We don't have enough weapons," added Chris Brose, chief strategy officer for the defense startup Anduril, which began as a border-surveillance automation company. "That pie needs to grow, and we want to help win a share of the growing pie."[56] The market knows, on some level, that these rockets only exist to *destroy* people and property, but Lockheed Martin investors don't have to factor in those costs. As far as capital is concerned, the best thing a substantial number of the world's people can do for accumulation is get hit with an expensive American missile. It's worth noting that the US Department of Defense is the world's largest institutional fossil fuel user: The military and its associated industrial complex have an annual carbon bootprint the same size as the forty-five least-emitting nations combined.[57]

Between the fossil fuel companies and the military-industrial complex, there is a significant capital bloc that has geared up to fight against the transition—a bloc that wants to transition global society to something other than the progressive path I've been trying to find in this book. It's not hard to guess how this bloc will express itself politically; we've encountered it as a problem in every chapter. It has fought and will continue to fight for its future by sabotaging transitional marketcraft, contesting public ownership, and killing communards. Though these capitalists have found a lot of bipartisan support among conservatives and liberals in carbon democracy—especially when it comes to belligerent foreign policy and expanding fossil fuel production—if progressives can secure one side, we can expect death capital to embrace the other. Concluding their extensive, globe-trotting study on what the authors describe as "the danger of fossil fascism," the Zetkin Collective writes that, to assess fossil fascism's likelihood, we have to ask two questions: "Will fossil capital defend itself to the bitter end?" and "Will it draw on the unique resources of the far right for the purpose?"[58] If the answer is anything north of maybe, and I believe it is, then we should expect the transition to be not just a metaphorical fight against emissions and complacency but also a literal fight against a strong social faction that seeks a worse world. To stretch the dancing pair metaphor, progressives of all types have no choice but to cut in.

CONCLUSION

THE PUZZLE

The planetary crisis is overwhelming, and it can become difficult to recall the strategies for progress or imagine how we might use them to surmount an obstacle the size of reality. My goal with this book has been to give an account of ideas that have gripped people and plans that have become actions (and vice versa), not to adjudicate between them. The three strategies are not potential keys to the crisis lock. Instead, they're more like puzzle pieces, with knobs and pockets that allow us to build something bigger yet still as solid: the Left. This conclusion's relatively modest ambition is to imagine some of these forms and sites of strategic connection, connections that might cobble up into a social form that is both much more powerful and much more flexible than its isolated parts. That, it seems to me, is the only realistic hope any of us has.

It's difficult to get the right amount of realism—we can no more countenance surrendering reality to ideals than principles to circumstances. My argument rests on three pieces of realism:

A. Marketcraft, public power, and communism are the Venn-diagram field of viable strategic action for progressives in the near term.
B. Partisans of one strategy will not persuade the others to give up and join them, not on a relevant time scale.
C. Partisans cannot blaze a successful path toward a better world on their own, outside the context of a larger left-wing strategy of strategies. Public power needs the radical threat; communists need bail money; marketcraft needs an organized working-class constituency, and so on.

The three strategies are distinct but entangled, a familiar kind of relation by now. Though there's nothing wrong with being, as I am, a partisan of one of the three, the full range of thought and planning is auspicious. We need at least one way out of the Oil-Value-Life chain, and all three strategies offer reasoned paths to a much better world, for the world. And yet I've already argued that we can't treat the planetary crisis as an occasion for a friendly contest between groups of partisans acting out their strategies in isolation. The challenge is to produce a model of how these strategies might click together in practice without requiring the unrealistic suppression or domination of one by another.

From a tactical point of view, the truth is that we have no shortage of right answers.* In one sense this surfeit should give us some relief, but in another sense it is deeply disturbing. If we have so many answers to our urgent planetary problems, then we can deduce the formidable strength of the force that is preventing us from devoting our time and

* "There is currently an efflorescence of proposals for how to decarbonize the economy, not only from the usual green capitalist policy wonks but also from agroecology enthusiasts, proponents of public banking and social housing, to those that tackle the logic of planned obsolescence and advocate for zero-waste production and consumption. I have never had so many conversations about the architecture of our electric grids, the relative contribution of distinct sectors to overall emissions, or the dilemmas of carbon taxes..." Thea Riofrancos, "Plan, Mood, Battlefield—Reflections on the Green New Deal," *Viewpoint,* May 16, 2019.

energy and other resources to enacting them at a sufficient pace, and that force is the voluminous dark matter of Value's mute compulsion. The only way for humans to unlock our capacity for invention and problem-solving is to confront and overcome that retarding force, first and foremost by winning a social-historical struggle that liberates our species to select and develop new values for ourselves. Now, how do we do that?

In 1962, a southern "white farm boy" named Kenneth Merryman sent a letter to the playwright Lorraine Hansberry—presumably as part of a school assignment—asking her what she thought about Martin Luther King Jr. and the "non-violent movement." The thirty-one-year-old Hansberry's response was immortalized in the posthumous biographical play *To Be Young, Gifted and Black*. "Like most of my generation I support and applaud Dr. King," she wrote.

> At the same time ... I have no illusion that it is enough. We believe that the world is *political* and that political *power,* in one form or another, will be the ultimate key to the liberation of American Negroes and, indeed, black folk throughout the world.... I think, then, that Negroes must concern themselves with *every single means* of struggle: legal, illegal, passive, active, violent and non-violent. That they must harass, debate, petition, give money to court struggles, sit-in, lie-down, strike, boycott, sing hymns, pray on steps and shoot from their windows when the racists come cruising through their communities.

"The *acceptance* of our present condition," she concludes, "is the *only form of extremism* which discredits us before our children."[1] What Hansberry describes is similar to a protocol called "diversity of tactics," in which elements in a coalition agree to adopt various rules of engagement *and* not to condemn one another to the public. This allows them to combine forces when it's useful—say, by building numbers for a protest march—while maintaining distance when it comes to whether to light stuff on fire. But Hansberry's phrase "must concern themselves with" suggests something more—that it's not enough to show up for a demonstration once in a while and that such a social struggle requires a conscious diversity of strategies.

Think of our situation this way: We face three paths for progress. People can make arguments, but there's no way to tell for sure which—if any—will continue to be viable as the weather changes, and we know the weather will change. Our best chance to find out appears to be to split up into thirds with the hope that some of us will make it, but no group of partisans has the numbers or resources to go it alone, regardless of aleatory conditions. No wonder so many caring people are sitting on the picnic benches of indecision at the trailhead of despair. Those who've strode ahead often look over their shoulders, turn around, and start jogging back to reevaluate. On some level, we know we need a higher level of agreement, and yet on another we know that the diversity of thought is valuable. The Left must walk down three strategic paths *at the same time,* and we have to do it all *together.* And we're already late.

What would it mean to do it, to immediately proceed down three paths at the same time? First, our metaphor requires different rules of physics, which is, thankfully, more realistic than it sounds. Though it's still hotly debated among experimenters, a faction of biologists believes that this kind of movement is the foundation for life as we know it.[2] In a "quantum walk," a particle or impulse of some kind advances down several paths toward its destination at once, collapsing into a single reality when it finds the fastest route—or if someone tries to take a peek at the process. The rogue biologist faction believes that's how photosynthesis works: Solar photons knock loose another piece in the cell, which finds its way to a pair of chlorophyll molecules with extraordinary efficiency, the kind you'd get from taking several paths at the same time. Though the volumes of mass involved are vanishingly small, and the time involved is vanishingly short, it's enough for a metaphor (and perhaps plant life in general). A quantum walk strategy of strategies means working to maintain a state of coherence in which we intentionally advance down a path that could be any of the three, putting ourselves in the right position as reality's potentials collapse into a particular unpredictable possibility. Why not? On the world-historical scale, we and this moment are vanishingly small, too.

Maybe marketcraft advances according to plan and capital is cajoled into accepting lower rates of profit, the stranding of fossil assets, and only some minor nationalization and a few exploding pipelines in the background; maybe we find our way to a confederation of governments, in which we agree to plan the world's metabolism on a democratic basis, beginning with climate reparations from the center to the periphery and an end to border walls; maybe workers of the oppressed nations (including Indigenous nations) launch a revolutionary joint dictatorship to smash global capital and impose agroecology; maybe people everywhere wrench control from their exploiters and call one another to celebrate and swap local specialty items; maybe public-power socialists win an election that prompts capitalist backlash and a full-blown revolutionary situation; maybe the ice caps will melt faster than forecasted, chilling the North Atlantic Current and triggering an abrupt climate shift, and we mostly have to worry about producing, organizing, and sharing life's primary mediations while we figure out what's next; maybe the fascists will move faster than we will and we'll be dragged into a fight no one wants on unbearable terms; maybe all of the above happens in an uneven manner around the globe; maybe there will be another world war. How do we prepare to meet this whole landscape of possibilities with quantum efficiency and speed? In the coming decades, how do we give ourselves a real shot? I devote what remains of the book to this modest question.

CHARACTERISTIC FEARS

As I've averred, psychological questions are mostly beyond the scope of this book. But I'd be remiss if I didn't acknowledge that one obstacle to progressive coherence is that our strategies are correlated with different temperaments. This is only logical, since the strategies call for and appeal to various orientations toward the status quo—it truly does take all kinds. But this is a serious weakness if we allow patterns of temperamental difference to harden into willful misunderstanding and distrust. In the interest of sustainable diversity, I want to recognize a few fears and anxieties I see as characteristic of (though hardly exclusive to) a number of strategic partisans. My hope is that addressing them quickly by name might allow us to control and even benefit from their effects rather than letting them fester into sources of disorganization.

Some fear **inefficiency,** which acquires a moral aspect as a critique of corruption or ideological partiality. We should write our programs and policies in a straightforward way, lest we lead ourselves astray with bad intuition or bend ourselves out of shape with discretion and exceptions. Although today's society is generally unfair, it's important to prefigure a better world by acting out the fairness we wish to see. That's all true, and yet for the left to succeed, it will require people to think politically first and foremost, and that can make for conflicting considerations. Marketcrafters can't let their dedication to prefigurative action isolate them from the rest of the left and their "outside" strategies, which, from a decarbonization perspective, say, seem comparatively roundabout and even dogmatically misguided. But the more we know about all the strategic visions, the easier it is to understand the logic behind one another's actions and maintain coherence—which is just another way to describe the future possibility that we will have been working on the same project all along.

Some fear **disorder,** which they understandably associate with social collapse and predation. We must ensure that we are fully prepared to replace any preexisting systems we plan to incapacitate or abolish lest we negligently make life worse for people and turn them against the left. Although there are many existing institutions and relations that we seek to overcome, it's important to maintain stability as we transition to a better social metabolism. That's all true, and yet for the left to succeed, it will require participants to take risks, including, at times, the risk of rapid change. Public-power advocates can't let their dedication to security isolate them from marketcraft and the communist partisans who seek hard and immediate changes—whether the abolition of coal plants or the abolition of the police. From a public-power perspective, both might appear as threats to the immediate well-being of working people. But the more we develop mutual understanding among partisans, the more prepared we are to organize for the better in a time of big, fast changes, which are happening whether the left initiates them or not.

Others fear **complicity,** which is a wholly justifiable response to a destructive metabolic order that generates and thrives on the vast majority's compelled cooperation. We must ensure that we aren't co-opted by the exploiter class lest we see our own work diverted in unfruitful directions or even turned against us. Although living in the capitalist world means making concessions simply to get access to our primary social-metabolic mediations, it's important to remain in constant conflict with the system that produces that disastrous situation in our lives in the first place. That's all true, and yet for the left to succeed, it will require participants to get their hands dirty working in coherence—if not alliance or coalition per se—with some private owners and politicians. Communists can't let their antisystemic commitments isolate them from marketcraft and public-power strategies that seek to make use of the material we've been dealt. Relinquishing the notion of political purity means a lot more precise analytical work regarding the divisions within our institutions, but that's the kind of work we need to do.

These fears aren't only reasonable, they're also usually helpful. But it's possible to apply reasonable fears unreasonably. For the left, absolutism is an internal threat, whether it's the outraged absolutism of revolutionaries or the staid absolutism of bureaucrats. For that reason, it's important not to treat our partisan analyses as morality tests. István Mészáros warns us that to identify what we understand as the situation's material dictates as moral imperatives is to lock ourselves "into the unbearably narrow confines of a permanent state of emergency."[3] There are exceptions—we can't be absolutist even in our rejection of absolutism—but without a fundamental coherence, there's no way to call for an emergency. If the alarm is always sounding, no one is listening for the ring.

POINTS OF COHERENCE

Those are some of the left's perforations, like the weak points between the fingers in a Kit-Kat bar. But what are the attracting forces that hold the partisans together, the material around which we might form one solid Kit-Kat fist? We can take a few of the essential overlaps for granted, such as support for decarbonization and the future well-being of the people who make up the working class. Following are a handful of precepts that could anchor coherence on the left.

The police are enemies of the people. I don't mean this in an abstract, theoretical way—though that is also the case—I mean it in the same literal way I've described other facts. The police beat communists in the street, threaten elected officials who seek accountability, and slurp up government budgets. More than the institution of the police, the social force of the police is the front line of the opposition to the left. We know it; they know it; we know they know it; they know we know they know it. Partisans have various roles to play, but whether we're in the streets trying to abolish or at city hall trying to defund, we can all agree that we have to defeat the police.

Women's collective self-liberation from thousands of years of patriarchal domination in all its forms around the world is one of the few existing social conflicts with the scale and potential to make a decisive progressive impact in the coming years. This fact has very different implications depending on where you live, but everyone on the left can hold it in common. That means we can't risk the disorienting consequences of opportunistic alliances with antifeminist reactionaries, whether in the shape of populist appeals to men as men or enemy-of-my-enemy friendships with antifeminist governments.

International solidarity is the only possible foundation for a left-wing victory of any substance in the coming years. There is nothing progressive about a union job manning a border wall, and the left must oppose militarist elements from all angles. Democratic global planning means treating the needs and interests and desires of all people and peoples as equally important, which would mean an end to the enduring colonial distribution of power and resources as well as an end to the production of de-development and sacrifice zones. Different peoples have different roles in that process, but it's a process we must share.

Building power is not just about kilowatt hours. Though a lot of the strategic focus is on decarbonizing this or abolishing that, everyone agrees there's going to be a lot of building going on. From food forests to solar panels to dense housing to bicycles to electricity cables to microorganisms to all sorts of stuff we don't even know about yet, we will have to build the plane as we struggle for control of the cockpit. As we look to influence production plans, we should keep the political struggle in mind—principally, the balance of class power at all levels. A deal to subsidize solar panels or electric cars could backfire if it isn't structured to protect labor's bargaining power in that sector. And producing tons of electric cars is probably not all that strategic if it further delays an inevitable social reckoning with a wasteful car-centered life that alienates working people from one another. Building power means being politically attentive as we pursue the constructive parts of our partisan strategies, thinking through the implications for the whole left field into the future.

Voting won't solve our problems, but those of us who can should still do it, and for a unified slate of candidates. Though voting is hardly the only tool we have, we should make the best use we can out of whatever rights we have under the current system, whether those are legal rights, to defend our safety; labor rights, to defend our jobs; environmental rights, to protect ecological systems; property and tribal rights, to protect our resources; speech rights, to push our ideas; and voting rights, to have our say, even between undesirable choices. For those who don't need to be persuaded to vote, this is a call to put forward candidates whom even people on the far left can bite their tongues and support rather than looking to hedge electoral bets with compensatory votes on the right. For the far left, it's a call to support the most viable choices you can.*

Fidelity to principle offers more than the cold comfort of a clean conscience. While our enemies understand opportunism and self-interest and short-term thinking, they do not understand the substance of our strategies and the principles that underlie them. As long as we hew to them, we have an advantage. It reminds me of a scene in the 1958 film *The Hidden Fortress,* which tells the story of a general's

* This was the tactical compromise between the French left and center blocs that successfully kept the right out of power during the 2024 elections: Each agreed to pull one of its candidates in the second-round race.

attempt to rescue a princess and escort her through enemy lines, along with two bumbling peasants and a bunch of gold. When the crew encounters an enslaved woman, the noble princess insists on trading their horses for her liberation. At first this slows their travel, but when a patrol comes upon the group, they can't recognize the outlaws: The authorities are looking for four people and some horses, not five people on foot. The protagonists are consistent substantially rather than in the superficial way their enemies understand. Similarly, our situation will change, and we'll be forced to continually reevaluate our choices and composition. But in the final instance, our shared principles should render us reliable and consistently legible to one another.

MODELS OF INTERACTION

Once we calm our various fears and embrace some points of coherence, we still have to find ways to interact successfully. It's not enough for partisans to go about their various strategies alone—if it were, we'd already be in a better situation. We need models for the ways in which people with these or similar strategies have worked together, less by way of compromise than via complementarity. The lowest common denominator, such as big marches and rallies, is insufficient. There's no need to line up in a popular front behind the left's most moderate demands; that would be far too risky. Instead we can look for imaginative designs, places where we can fill in the overlaps in our Venn diagram. There are plenty.

Marketcraft and public power: Of the four overlaps, this is probably the most well developed, both in theory and in practice. The People's Republic of China has rapidly come to lead the world in the production of decarbonization tools—from solar panels to batteries to pumped hydropower to trees—via an overlapping marketcraft and public-power program that is the envy of every government in the world, even if they can't say so out loud. State policymakers set the direction for the economy using a variety of tools, including sectoral planning, layers of public finance, state-owned enterprises, cheap land provision, a strong bureaucracy, capital controls, and the occasional disappearance of prominent businessmen who get too big for their britches. Despite the Western depictions of the Chinese economy as command-based, those are all ways of crafting markets.

"The fundamental advantage of a market economy is not an improvement in individual decision-making per se. No-one can precisely predict an uncertain future, and in a free market there will be much more failures than successes," explains economist Lan Xiaohuan, describing the government outlook. "The advantage of a free market is that it ensures competition and through a process of trial and error, only the fittest or best-adapted companies survive. Therefore, industrial policies that are designed to support competition, rather than supporting specific companies, tend to be more effective at stimulating development."[4] The "free" market operates as a virtual computer inside the mainframe of state policy, and the program's values are subject to deliberation. As Xiaohuan puts it regarding state infrastructure investment, "It's better to calculate not only the project-level financial returns (which may be low), but also the overall benefits brought to the economy and to society (which may be much higher). However, there is no consensus on how to evaluate these social and macroeconomic benefits, so the values of many infrastructural projects are always open to debate."[5] This goes for the economy in general.

For example, scholar Wei Zhang has called for a shift in the country's approach to health care, away from market commoditization of medicine and toward a focus on the social determinants of health—quality, not quantity. "As a socialist-oriented country with a relatively large state sector and strong state capacity for macroeconomic planning and regulation," he writes, "China should take full advantage of its potential to mobilize resources and power to build a health-enhancing society."[6] Acknowledging the decisive importance of non-Value values and society's prerogative to enforce them on economic actors is a vital accomplishment. There's no need to repeat what I've already written, but it's worth acknowledging that the PRC is likely to be the dominant model for this kind of cooperation in the near future.

Still, China's is only one model. Malawian economist Thandika Mkandawire conceived of a model for an African "developmental state" that fends off predatory international capital by creating tame Indigenous capitalists who are directed to invest in the country rather than extract from it.[7] More recently, theorists have adapted Mkandawire's thinking for the era of decarbonization, imagining national green developmental blocs undergirded by a new "democratic global economic and monetary order" designed to "reduce economic inequalities between countries and between those bearing the brunt of climate change and climate injustice."[8] Here development is driven by domestic demand rather than the export market, making it easier in theory to secure political support and participation among the masses. Developing countries can leapfrog the West technologically and ecologically, jumping from bicycle to e-bike without the need to produce a car per person in between. Politically, this isn't all that different from what Roosevelt accomplished with the TVA, building a coalition between workers cut off from industrial development and regional capitalists who would benefit from electrification.

National marketcraft also eases the way for workers who seek to build their collective power independent of the state. The state can do this directly, but it must impose conditions: Firms seeking subsidies or contracts would be compelled to deal with organized labor. For example, pro-union provisions in the Inflation Reduction Act helped allow workers at an electric school bus factory in Georgia to push back against their bosses and seek representation by United Steelworkers.[9]

By supporting strategic industries, policymakers can improve labor's situation.

Finally, there is insurgent public power from below, which can rope in some traitors from the opposition capitalist class by pitting them against one another. This doesn't require capitalists to work against their own interests—at least not their narrow immediate interests. As long as green capital is sufficiently distinct from fossil capital, it has an interest in supporting its competition's workers against their bosses. We can imagine a cabal of clean energy investors who seek to lower the cost of labor actions for workers in the oil and gas industry by maintaining a large strike fund, for example. Together, this bloc could twist the market away from fossil fuel capital and divide the spoils, with labor insisting on an improved share from the new (and comparatively weak) bosses.

Public power and communism: In this overlap, we see the official labor movement assume a lead role for the whole class not by definition but through its actual dedication to the task. Social leadership is earned, not derived. In Brazil, for example, the trade unions have positioned themselves as the anchor for a variegated left-wing civil society forced to do battle with the right rather than as a special interest group. Unions led a 2017 general strike against the right-wing government's austerity measures and rallied to the support of imprisoned once-and-future president Luiz Inácio Lula da Silva, himself of course a former union leader.[10] But as filmmaker and anthropologist André Singer writes, analyzing the results of the 2022 election that returned Lula and the Partido dos Trabalhadores to power, it was "sub-proletarian" support from "seasonal agricultural labourers, street vendors, informal security guards, the off-the-books employees of small manufacturers, undocumented domestic workers, and so on—who find themselves 'deprived of the minimal prerequisites for participating in class struggle'"—that made the difference, along with "women, blacks, indigenous peoples and LGBTQIA+" (overlapping categories, no doubt).[11] By maintaining coherence with the unofficial labor movement, the official labor movement helped lead the working class back into power. Lula has worked to repay the favor, positioning himself as supporting legal abortion and forming a commando squad under the ministry for the environment—*not* under the right-wing military—to protect Indigenous Yanomami territory from artisanal miners.[12]

In the United States, where trade unions do not have much political representation or popular power, workers can still use their resources and numbers to provide a foundation for those left out of power. The International Longshore and Warehouse Union (ILWU), which operates primarily on the Pacific coasts, has held on to its radical heritage while other trade unions have turned toward a conciliatory model. The ILWU (especially Local 10, in the Bay Area) has allegedly coordinated with the radical left, offering a willingness to walk off the job (in the name of workplace safety) in response to—in aid of—street demonstrations and blockades. The ILWU has shut down ports not only for its own wage actions but also as part of racial justice and antiwar protests. The union has enforced boycotts against right-wing governments in South Africa and El Salvador, using its place at the hinge of global capital flows to assert international solidarity and working-class self-determination.* In 1999, the ILWU was one of unions that held down the massive protests that forced an early end to the World Trade Organization meeting in Seattle, standing back-to-back with insurrectionary anarchists against police tear gas. As one member recalled, at a crucial moment the ILWU leadership on the ground "agreed to continue into the fray, rather than ignore the desperate urging of the other protesters who were in a standoff with the Seattle police at several

* For a history of ILWU internationalism, see Peter Cole, *Dockworker Power: Race and Activism in Durban and the San Francisco Bay Area* (University of Illinois Press, 2018).

barricade sites."[13] When the city insisted on holding masses of arrested protesters, the ILWU threatened to shut down the Port of Seattle, forcing their release. Though refusing to handle Salvadoran coffee sounds like clean-hands conscious consumerism—perhaps even an example of fear of complicity gone too far—with the union's help it becomes genuine political struggle.

In most national situations, where the workers remain outside of power, the official labor movement is most likely the only institutional actor with the social credibility, resources, and discipline to begin arming the left. This is a crucial role, considering the mounting pressure from fascists—both military and paramilitary. If the left can't maintain parity in this field, it effectively encourages the right to seek bullet-based solutions to public-power challenges. Witness the recent assassination of up-and-coming young left-wing politicians Marielle Franco (Brazil, 2018), Hevrin Khalaf (Rojava, 2019), and Brigitte García (Ecuador, 2024)—not to mention the 2011 attack on the Workers' Youth League camp in Utøya, Norway, or the 2012 massacre of striking members of the National Union of Mineworkers in Marikana, South Africa, or the 2014 mass disappearing of Mexican students associated with the Coordinadora Nacional de Trabajadores de la Educación, a teachers union, in Iguala, each of which left dozens of *leaders* murdered. The workers' militia is about more than being ready with a counterforce when the bosses get tired of contract negotiations, though that is an important aspect, too; it's also about leading the whole class. As one underground resistance journal in occupied France asserted in 1944, under the article title "How Do You Form a Workers Militia?," "The Workers Militia enters into contact with factory and neighborhood cells and attempts to constitute them if they don't yet exist. It makes contact with the workers parties, the illegal unions, the [communist militias,]" and women's organizations.[14] When the official labor movement recognizes itself as an important part of a larger whole, it makes left coherence possible.

Communists, as I've employed the term, can't ultimately permit violence monopolists. They can't countenance any everlasting political organ that seeks to centralize control of the ecosocial metabolism. That seems to conflict with the state-centric public-power strategy. It wasn't until I considered public power in its most literal form that the state's transitioning to something uncoercive over time in a free society made sense to me. In a metabolism characterized by FETE, operating large-scale electricity generation, transmission, and storage facilities need not be a form of social domination. Public power (political) becomes public power (electrical), one mere set of nodes on the world's nonhierarchical metabolic web. Important nodes, no doubt, but hardly dominating or representative, especially if coupled semi-redundantly with a more widely distributed energy system. It reminds me of a sketch by the comedy group the Whitest Kids U'Know about nuclear power-plant workers following a successful anarchist revolution. They're happy with anarchy *and* happy to keep doing their jobs cooling the reactor, but they want to make sure they'll still get food. The joke is that the new society has to keep inventing the police and political leaders in order to verify that the power-plant workers and only the power-plant workers—plus the police and politicians themselves, it turns out—are getting fed for free. But by severing the "from" and "to" of metabolic provision, communists solve the problem, and efficiently, without the drag of feeding people to oppress us. Minus the police, the military, and politicians, the state could wither into its particular actually useful functions. This needn't hinge on a single revolutionary moment; today's leaders can help transform the state in this direction in advance. One perhaps surprising example comes from North Korea, where Kim Jong-un's government seems to be quietly converting military airfields into giant vegetable farm greenhouse complexes.[15]

Organized workers accomplish a similar withering-to-usefulness of the capitalist production system if they plan well from below. When ownership shut down the unionized Republic Windows and Doors factory in Chicago, the workers disagreed, occupying the factory. They formed a collective and, with union help, bought their own factory to run as a cooperative, where they now specialize in producing high-quality energy-efficient windows.[16] Or consider the Italian workers at a closing auto parts factory near Florence who have occupied their facility for years in a broad alliance with local left-wing movements, "from feminists to green causes."[17] In the name of "reindustrialization from below," the workers have planned to transition the factory to a facility that produces solar panels and frilless cargo bikes; they are raising the capital by selling €100 shares in the new firm to community supporters.[18]

And yet there's something missing in these stories, which don't quite get to the boundary-smashing transformation of the individual's relation to society that communists strive for. Workers' control of work is central to public power, but that's not the same as the overcoming of work itself as the center of life. In one of my favorite examples of public power–communism overlap, workers at a coal-fired West Virginia power plant have taken it upon themselves to cultivate chestnut trees at their worksite. Led by tech services supervisor and chestnut enthusiast Jon Durbin, who harvested the seeds and germinated them in his fridge, Harrison Power Station employees planted and fenced seven trees on the grounds. "It's just sort of a hobby—something I like to do," Durbin told the local news. "I wanted to see if I could do it. It's environmentally friendly in a bunch of different ways. They do absorb carbon. They give off oxygen. There are a lot of deer here on site. It's food for them."[19] The company has been happy to publicize the efforts, but it's obvious that, if they survive to maturity, the chestnut trees will long outlast the coal plant. In one small way, Durbin and his coworkers are planning the transition and making it happen, providing for the future needs of society (and deer) according to their own thoughtfulness, imagination, and initiative.

Communism and marketcraft: Of the four models of interaction I describe, the one between communism and marketcraft is probably the least well developed. If communists are collaborating with capitalists, doesn't that mean there has to be some confusion or hypocrisy going on by definition? Not necessarily. Decades after his death, Vladimir Lenin was credited with saying that when the communists are ready to hang the capitalists, the capitalists will bid against each other for the rope contract. Or, as it's sometimes phrased, "The last capitalist we hang will be the one who sells us the rope." If the fossil capitalists are willing to produce the world to death because the markets goofed on weighing the pollution variables, then why wouldn't a green capitalist or two be happy to sneak up behind the fossil fuel industry and slit its throat if there were something in it for them? After all, that's what the oil industry has done to its competitors for a century. Capitalist class betrayal is now a potential growth industry, and not for the first time.

To find moments of common purpose, communists don't have to rely on the moral fiber of individual ideologically committed left-wing capitalists—though the truth only whispered on the communist left is that such people exist and have existed for as long as the capitalist mode of production has, individuals being strange and unpredictable. Instead, it's a bet that capitalists are able to make mistakes, that it's possible for class enemies to be led into misperceiving their interests, something we already know they're capable of based on the way they're undermining the environmental premise for their own future profitability. That's the idea behind getting capitalists to bid for the rope contract. Factions of capital routinely find themselves in serious long-term disputes and zero-sum situations, and communists can take advantage of those divisions for their own purposes. Slavery abolitionists in the United Kingdom and the United States agitated for slavery-free goods, joining with producers who resented the unfair advantage that unpaid labor bestowed on their competitors. Even John Brown and his insurrectionary family depended on the support of sympathetic capitalists, who came to feel involved in the struggle and whose fidelity to the radical abolitionist mission continued after Brown's death.

It's difficult to imagine a "marketcraft from below" that's distinct from individual conscious consumerism and worker co-ops, but communists imagine this all the time. "Our movements must contend with the structures in place in order to dismantle the weapons they use against our communities, and simultaneously build new ways of surviving that are based in our principles of liberation and collective self-determination," writes Dean Spade at the end of his short book *Mutual Aid: Building Solidarity During This Crisis (and the Next).* "We must imagine and build ways of eating, communicating, sheltering, moving, healing, and caring for each other that are not profit-centered, hierarchical, and destructive to our planet."[20] These mutual aid efforts are, at worst, charity campaigns, but at best they build left-wing power, too, providing people access to their primary metabolic mediations under FETE in defiance of the Value regime. And there's no need to stop at primary mediations—"No one turned away for lack of funds" (NOTAFLOF) is the communist protocol for access to community events, whether a fundraising dinner or a punk concert. Communists shouldn't be afraid to solidify these projects, even if that means asking people to exchange money, if they have it, for goods and/or services that the communists organize to produce. And though they may find them aesthetically objectionable, traditional marketcrafters can find coherence with these activities by thinking about the progressive pressure they can exert on the existing system and creating channels in that direction. That means, for example, coming to the defense of building squats and homeless encampments as they reveal the need for more, better housing and put reformers in a better position to seek it.*

Capitalists struggling with one another won't end the Value system, and the history of modernity is the history of new bosses. But if we have any choice in our enemies, we're better off contending with a regional chestnut cartel than mostly fossil fuel agricultural conglomerates such as Cargill. Weak capitalists are better capitalists as far as we're concerned, and communists can support antitrust and other efforts to undermine the power of existing firms. Some communist partisans have sought to bring grassroots pressure to bear against rule-skirting tech companies under the banner of neo-Luddism, using fiery property destruction to stall the automated car of capitalist knavery while well-meaning regulators catch up.[21] Here, too, communist rage in the street is the progressive bureaucrat's friend, providing clear examples of sympathetic public commentary—as long as the bureaucrats are willing to work through their visceral fear of disorder.† Though the focus has been on protesting the deterioration of labor relations, we can imagine neo-Luddism expanding to take on military, surveillance, and fossil fuel tech in everyday life as part of a generalized political movement.

All in together: As I've said, it's not realistic to expect any set of partisans to give up their work in order to unite under a different strategy, even if we could be sure that's a good idea, which we can't. A

* For more on what the author calls "urban climate insurgency," see Ashley Dawson, *Environmentalism from Below: How Global People's Movements Are Leading the Fight for Our Planet* (Haymarket Books, 2024), 73–110.

† "Labor strikes, rent strikes, strategic defaults, urban riots, occupations of public space, and squatting can all have an acute effect on the value of private wealth and the calculation of public spending priorities." Melinda Cooper, *Counterrevolution: Extravagance and Austerity in Public Finance* (Zone Books, 2024), 383.

metastrategy of coherence holds everyone together to a lesser standard. Spatially, the entire Venn diagram fits in a circle of coherence, or at least that's the idea: The Left concerns itself with the whole field. But there in the middle, no matter how pinpoint small, is the three-way overlap. I don't think there's any sort of magic key hidden there, but it's worth trying to understand, as is every other part of the field.

One way to understand the central overlap is as an emergency siren. In certain moments, everyone's work is aligned because there really is no choice. In moments of acute disaster, for example, everyone ought to focus on organizing and sharing primary mediations such as food, water, and shelter. Different partisans have different ideas about how best to accomplish that, but the urgency and shared values should tighten the scope of our practical activities in emergency scenarios. Marketcrafters can understand "looting" cases of water from Walmart; communists can understand the arbitrary authority involved in rationing scarce supplies; public-power advocates can understand taking action without waiting for official permission.

Disasters and emergencies aren't always natural, though: The rise of fascists in the public sphere is just as much cause for alarm as a hurricane. In 2017, buoyed by the election of Donald Trump, fascists in the United States, gathered under the new label alt-right, began organizing publicly, forming open organizations, marching in the street, and developing relationships with elected officials. With novel levels of support among the police and a vicious eagerness to escalate, the alt-right looked ready to overwhelm the small number of anarchist and communist antifascists who have kept the creeps under control in the past. But perhaps aided by partisan anti-Trump animosity, more or less the whole left showed up to face down the right, and the police had no choice but to protect the outnumbered fascists from the public.[22] Those on the left whose distaste for disorder might have kept them at home in the past, who normally believe in fighting this kind of thing in the marketplace of ideas, were able to hear the alarm and understand its significance, even though anarchist rascals raised it.

A similar coalition has animated the Stop Cop City campaign in Atlanta, where a planned police training facility threatens the Weelaunee Forest. Radical forest defenders have led the effort from the beginning, putting their bodies on the line to obstruct progress—most notably in the case of Manuel Esteban Paez "Tortuguita" Terán, who was shot and killed by the authorities in January of 2023. While these radicals have turned Atlanta into a national center of politicized property destruction by attacking construction equipment, that hasn't stopped them from expanding their base to include local schoolchildren and the American Friends Service Committee, founded by Quakers. Stop Cop City has fought the complex on every level, from alleged "terrorist" arson to a city referendum to intracapitalist pressure to community festivals and tree planting. But even if it's the official lane where the campaign ostensibly succeeds and the formal organizations get credit, we shouldn't ignore the overlap's importance. As two participant-observers write, if you really want to know what's going on with Stop Cop City, "ask an anarchist": "They all know who painted the banners, who printed the zines, who organized the inaugural info night. Who barbecued the jackfruit, who hauled in the speakers, who gave the movement its slogans and myths and indefatigable energy. Who got neighbors and strangers together to do something more than post about it. Who transformed concerned citizens into forest defenders."[23] Radical verve is an important resource, and no part of the left can forsake it except at its own peril; finding ways to embrace this energy can focus the lens of left coherence to a dangerously hot point of light.

The focused left can do more than oppose. In the 1980s, the left organized to support the revolutionary left-wing Sandinista government in Nicaragua, in spite of American sanctions meant to destabilize the regime. Using a loophole in the law, some food co-op workers in Boston began importing Nicaraguan coffee beans en masse from a Dutch roaster and selling them through an informal national network of left-wing coffee shops—an odd form of countermarketcraft.[24] With this work, they undermined the Cold War effort to increase the exploitation rate in Latin America, partly by imposing cultural pressure on members of the American left to use their exceptional coffee consumption rates to support independent shops rather than the burgeoning chains that took advantage of new lower wages in coffee-producing countries. The same network proved valuable when it came to the labor-led boycott of Salvadoran coffee, which helped pressure the right-wing government to halt its murderous counterinsurgency campaign. Often derided as "consumer politics," these campaigns in reality connected coffee workers, national governments, the coffee market, and anarchist teenagers who loitered outside Starbucks jeering at the patrons. The chains may have won the day as far as today's American customers are concerned, but the international solidarity campaigns succeeded.

What about the few nations where the left is in power? How does the broad left contend with the contradictions between state dictates and communal practice? I've mentioned Venezuela, where government policy makes a place for communes, ironically, in the market. One popular example is the Luisa Cáceres Commune, which was one of nine communes awarded garbage-collecting contracts by the city. Public power using marketcraft to build communes sounds relatively down-to-earth and, well, legal. Hardly revolutionary. But as we know, communes are not mere cooperatives. After building up local support for its work, the Luisa Cáceres Commune appropriated a subsidized neighborhood grocery store, which was being transferred into private hands. "Good afternoon," the communards told the store manager. "Please give us the keys and your phone. This Mercal is now in the hands of the commune."[25] The communards cleaned, fixed up, and expanded the grocery, creating a location not only for food distribution but also for community gathering and education.* Within a coherent left, planning from above can seed planning from below.

We should understand these as real attempts to deal with the current situation rather than as low-stakes ideological experiments. Facing economic isolation in the '90s, the Cuban people developed the world's most advanced system of urban agroecology. When the urban population found itself literally starving after the Soviet Union's collapse—the USSR had been Cuba's main trading partner and food and fuel provider—the Cuban government accelerated the development and deployment of *organopónicos,* an urban farming technique featuring long raised cultivation beds filled with soil and natural fertilizers, and legalized land occupations by producers.† Spearheading the initiative was Moisés Sio Wong, a Cuban general and descendant of immigrant Chinese farmworkers who once introduced vegetable plots to the island's degraded colonial foodscape. By backing agroecology with public power, Cuba has secured the *organopónicos* as an element of city life. Many are located near dense housing areas and include vending stands where consumers can buy directly; they distribute produce from the countryside as well, being farm and farmers market in one.[26] And the communal character extends beyond the circuit of production and consumption: Economist Sinan Koont, who's made an extensive study of Cuban urban agriculture, describes one *organopónico* in Havana called INRE-1 that not only supplies local schools with fresh

* For detailed accounts of Venezuela's experimental economic forms, see Dario Azzellini, *Communes and Workers' Control in Venezuela* (Haymarket Books, 2017), and Chris Gilbert, *Commune or Nothing!: Venezuela's Communal Movement and Its Socialist Project* (Monthly Review Press, 2023).

† For more on how the Soviet collapse and the increasing integration of China into the unipolar capitalist bloc affected fossil fuel food production in satellite states, see Zhun Xu, "Industrial Agriculture: Lessons from North Korea," *Monthly Review,* March 1, 2024.

vegetables for their daily lunches but has also integrated itself into the life of one particular school: "Its workers have more or less adopted the school and taken it under their urban agricultural wings," he writes. "They constructed a small organopónico on school grounds where circle of interest students [sic] engage in actual cultivation... In addition, INRE-1 has helped the students construct a garden called Ruta Martiana (Martí Lane) containing all of the Cuban flora José Martí cited and described in his journal."[27] The state delegates a market function to the workers, and the workers expand their functions in the process of developing new communal social-metabolic relations.

DISASTER COUNCILS

There are a few existing models for progress on the left. The most famous is undoubtedly Martin Luther King Jr.'s formulation that "the arc of the moral universe is long, but it bends toward justice," to which he sometimes added: "With this faith we will be able to hew out of the mountain of despair a stone of hope. With this faith we will be able to transform the jangling discords of our nation into a beautiful symphony of brotherhood."[28] This reassures people that they're in a winnable fight, and events have verified it time and again throughout modern history.

Philosopher Walter Benjamin offered a different vision in his theses on the philosophy of history, in which he used Paul Klee's painting *Angelus Novus* to model the "angel of history" whose "face is turned toward the past":

> Where we perceive a chain of events, he sees one single catastrophe which keeps piling wreckage upon wreckage and hurls it in front of his feet. The angel would like to stay, awaken the dead, and make whole what has been smashed. But a storm is

> blowing from Paradise; it has got caught in his wings with such violence that the angel can no longer close them. This storm irresistibly propels him into the future to which his back is turned, while the pile of debris before him grows skyward. This storm is what we call progress.[29]

Here there is nothing reassuring about progress except that things keep changing. By contrast, revolution is a tiger's leap into history's open air—an explosive movement into the unknown, without guarantees. There's something verifiable about this version, too, with its jumps of revolutionary change that surprise even the participants. Rome wasn't built in a day, but it fell in one—both are kinds of historical progress.

How, then, to model the struggle for progress in light of the present moment of planetary crisis? We don't have time for a long arc or blank destinations. We need an idea of where we're going, and we need to leap there. There is also the opposition to consider, because we're not the only actors on the historical field.* It's necessary to acknowledge the possibility of failure, of defeat. Our conjuncture is a complex one, with a number of parts moving in various progressive ways at the same time. We can use another painting: instead of *Angelus Novus,* Henri Rousseau's 1908 canvas *Les joueurs de football.*

Decidedly less epic than Klee's, Rousseau's piece has more movement. It depicts an early two-on-two game of "rugby football" played in a forest clearing, frozen in the moment one team tries to complete

---

* As an analytical matter, I unfortunately agree with geographer Phil A. Neel when he writes that, faced with the strategic opening that comes with a breakdown in the existing social metabolism, "the far right can mobilize its connections to police and military bureaucracies as well as the criminal and mercenary underworld in order to assemble and deploy its resources much faster than its largely undisciplined, untrained leftist opponents." Phil A. Neel, *Hinterland: America's New Landscape of Class and Conflict* (Reaktion Books, 2018), 49.

a long pass. The passer is in the back, having let the nutshell ball fly. His teammate is running forward, directly toward the viewer, arms stretched heavenward to receive. Behind him is a po-faced defender, his right arm stretched out in a punch that just touches the receiver's striped jersey. If the receiver slows even a half step to look for the ball, he'll be hit. He has no choice but to run his route, extend his arms, and expect that the pass will be there. The passer, meanwhile, has to throw not to the spot where the receiver is but to the spot where he's going, which means he has to trust the plan *and* adjust it according to the moment's particularities. The arc of the pass can only be so long, and everyone else in the painting watches the receiver; the receiver's determined eyes lock on the viewer as he hurtles himself toward a future he can't see but might still catch.

Perhaps it's too much to ask of the reader to combine our metaphors, but the left faces bigger challenges. If our quantum walk is a football play, then maintaining coherence means keeping the whole team on the same playing field. To accomplish that, we need to build relationships between partisans. Out of strong political communities we can create a broader culture that yearns to explode into a whole new metabolism, one way and another. I don't mean an alternative subculture; I mean a true counterculture, the practical collective imposition of new values that break the Oil-Value-Life chain and put humanity in position to improve the world for the world. Organs of left social coherence can take any number of forms, from sports leagues to community dinners to coffee shops to parties (both kinds) to farmers markets to language classes to international delegations to reading groups to tenant unions to import-export businesses to street demonstrations to radio shows to electoral campaigns to movie screenings to pool halls to workers' centers to harvest festivals to legal-defense fundraisers to religious congregations to bars to prisoner support meetings to antifascist militias to bingo nights, but it doesn't happen automatically. We can cultivate the conditions for a political culture, one from within which partisans can more coherently pursue their range of strategies. It's worth trying.

In this spirit, I end the book with a firm proposal: The left should lead the formation of community disaster councils. If there's one thing we know about the near future, it's that it will be if not disastrous then at least disasterful. Every place will see the world's conflicts erupt in particular ways, local crystallizations of the planetary crisis. Heat waves, fires, storms, floods, disease outbreaks, civil conflict, algal blooms, drought, energy shortages: The smooth skeletal sphere of capitalist metabolism is cracking. So far, capital's plan is to spackle those cracks with rubble made from the destroyed lives of the least fortunate—a plentiful and renewable resource—but the authorities accomplish this in large part by refusing to lead society to adequately plan for these disasters, preferring to leave everyone to themselves and the hindmost to the devil. The Value system's solution is clear; the left is obligated to construct and impose an alternative.

Disaster councils would bring people (and their various institutions and groups) together at the community level—whether a community of neighbors or an international community of climatologists—to plan disaster responses guided by progressive values rather than money Value. That means studying which disasters are likely to affect our communities and in what ways, then organizing people and institutions to ready a response that prioritizes the well-being of all people above the preservation and orderly accumulation of value. Maybe that's planning to occupy air-conditioned buildings as emergency cooling zones—or planning to occupy heated buildings as emergency warming zones. Maybe it's readying for the appropriation and distribution of food and water from grocery stores. Maybe it's rallying votes for a library funding referendum or supporters for a picket line at a dangerously overheated workplace. Maybe it's finding hotels to use as overnight shelters for displaced people. For the geographically dispersed communities, perhaps it will be more about channeling outside resources and help to local groups. Regardless, a disaster council can only acquire substance and legitimacy through action.

Disaster councils would allow people to come together without leaving the places they call home. Rather, the form requires participants to make use of their existing relationships, sprangling the council's activity through their families, friendships, workplaces, associations, what have you. On such a council, marketcrafting politicians could meet unionized workers could meet communist miscreants as equals, all of us patching together a new world the best ways we can figure out how and preparing to fight for it, together. That's what's left, and that's worth doing.

NOTES

Introduction

1 Malcolm Harris, "Shell Is Looking Forward," *New York,* March 3, 2020.
2 Dan Welsby et al., "Unextractable Fossil Fuels in a 1.5 °C World," *Nature* 597 (September 8, 2021): 230–34.
3 "Proved Reserves and Proved Undeveloped Reserves," 2022 Shell Global annual report, https://reports.shell.com/annual-report/2022/strategic-report/segments/oil-and-gas-information/reserves.html.
4 Gabriel Malek et al., "Transferred Emissions: How Risks in Oil and Gas M&A Could Hamper the Energy Transition," Environmental Defense Fund; see also Jack Arnold et al., "Transferred Emissions Are Still Emissions: Why Fossil Fuel Asset Sales Need Enhanced Transparency and Carbon Accounting," Columbia Center on Sustainable Investment, May 2023.
5 Hiroko Tabuchi, "Oil Giants Sell Dirty Wells to Buyers with Looser Climate Goals, Study Finds," *New York Times,* May 10, 2022.
6 "18% of ExxonMobil and Chevron Shareholders Support Accurate Reporting on Emissions Targets," As You Sow press release, May 31, 2023.
7 "Shell Backtracks on Climate Targets," Follow This, March 14, 2024; "Organizing Support for Climate Resolution at Shell Signals Shareholder Dissent over Climate Retreat," Follow This, May 21, 2024.
8 "Enterprising Child Saves $54 to Buy Barrel of Oil," *The Onion,* October 20, 2004.
9 Robinson Meyer, "Carbon Tax, Beloved Policy to Fix Climate Change, Is Dead at 47," *The Atlantic,* July 20, 2021.

10 Zoe Todd, "Fish, Kin and Hope: Tending to Water Violations in *amiskwaciwâskahikan* and Treaty Six Territory," *Afterall* 43 (Spring/Summer 2017): 103–7.
11 Todd, "Fish, Kin and Hope," 106.
12 Matthew Huber, *Climate Change as Class War: Building Socialism on a Warming Planet* (Verso Books, 2022), 55.
13 Simon Pirani, *Burning Up: A Global History of Fossil Fuel Consumption* (Pluto Press, 2018), 2.
14 Brett Clark and John Bellamy Foster, "The Dialectic of Social and Ecological Metabolism: Marx, Mészáros, and the Absolute Limits of Capital," *Socialism and Democracy* 24, no. 2 (2010): 124–38.
15 István Mészáros, *Beyond Capital: Toward a Theory of Transition* (Monthly Review Press, 2010), 44–45.
16 Mészáros, *Beyond Capital.*
17 Mészáros, *Beyond Capital,* 116.
18 "U.S. Imports of Goods by Customs Basis from Chad (IMP7560)," Federal Reserve Economic Data, accessed August 16, 2023.
19 "Chad Export Project: Project Update No. 32," Esso Exploration and Production Chad Inc., midyear report 2012.
20 Pat Davis Szymczak, "Chad Nationalizes Doba Project, Derails Exxon's Asset Sale to Savannah Energy," *Journal of Petroleum Technology,* March 28, 2023.
21 "Chad: AFCW3 Growth and Diversification Leveraging Export Diversification to Foster Growth," World Bank, May 30, 2019.
22 Xinshen Diao et al., "Africa's Manufacturing Puzzle: Evidence from Tanzanian and Ethiopian Firms," National Bureau of Economic Research working paper 28344 (January 2021), 5.
23 Dani Rodrik, "Premature Deindustrialization," National Bureau of Economic Research working paper 20935 (February 2015); Aaron Benanav, *Automation and the Future of Work* (Verso Books, 2020), 21. See also Phillip Neel, "Broken Circle: Premature Deindustrialization, Chinese Capital Exports, and the Stumbling Development of New Territorial Industrial Complexes," *International Labor and Working-Class History* 102 (Fall 2022): 94–123.
24 Acha Leke, Peter Gaius-Obaseki, and Oliver Onyekweli, "The Future of African Oil and Gas: Positioning for the Energy Transition," McKinsey & Company, June 8, 2022.
25 Fiona Harvey, "World's Biggest Economies Pumping Billions into Fossil Fuels in Poor Nations," *The Guardian,* April 9, 2024.
26 Mike Davis, "Who Will Build the Ark?," chap. 1 in *Who Will Build the Ark: Debates on Climate Strategy from New Left Review,* ed. Benjamin Kunkel and Lola Seaton (Verso Books, 2023), 18.

Chapter 1: Marketcraft

1 Alessio Terzi, *Growth for Good: Reshaping Capitalism to Save Humanity from Climate Catastrophe* (Harvard University Press, 2022), 11.
2 Thomas L. Friedman, "The Power of Green," *New York Times,* April 15, 2007.
3 Friedman, "The Power of Green."
4 Brian Deese, "The New Climate Law Is Working. Clean Energy Investments Are Soaring," *New York Times,* May 30, 2023.
5 "Five Years' Worth of Clean Energy Investments Announced in Less Than Nine Months," American Clean Power, April 17, 2023.
6 Karen E. Fields and Barbara J. Fields, *Racecraft: The Soul of Inequality in American Life* (Verso Books, 2012).
7 Steven K. Vogel, *Marketcraft: How Governments Make Markets Work* (Oxford University Press, 2018), 1.
8 Joe Stephens and Carol D. Leonnig, "Solyndra: Politics Infused Obama Energy Programs," *Washington Post,* December 25, 2011.
9 "Recognizing the Duty of the Federal Government to Create a Green New Deal," H. Res. 109 (2019), https://www.congress.gov/116/bills/hres109/BILLS-116hres109ih.pdf.
10 Stephanie Kelton, *The Deficit Myth: Modern Monetary Theory and the Birth of the People's Economy* (PublicAffairs, 2020).
11 Erin McCormick, "Rise of the Yimbys: The Angry Millennials with a Radical Housing Solution," *The Guardian,* October 2, 2017.
12 Isabella Weber, *How China Escaped Shock Therapy: The Market Reform Debate* (Routledge, 2021).
13 Xudong An, Stuart A. Gabriel, and Nitzan Tzur-Ilan, "More Than Shelter: The Effects of Rental Eviction Moratoria on Household Well-Being," Federal Reserve Bank of Philadelphia working paper, March 10, 2021; Ioana Marinescu, Daphné Skandalis, and Daniel Zhao, "The Impact of the Federal Pandemic Unemployment Compensation on Job Search and Vacancy Creation," *Journal of Public Economics* 200 (August 2021).
14 Elizabeth Shogren, "Why Dawn Is the Bird Cleaner of Choice in Oil Spills," *Morning Edition,* June 22, 2010.
15 Andrew Adam Newman, "Tough on Crude Oil, Soft on Ducklings," *New York Times,* September 24, 2009.
16 Rod Nickel and Nia Williams, "Oil Companies Cautious About Drilling as Energy Transition Looms," Reuters, September 20, 2023.
17 Chris Hughes and Peter Spiegler, "Marketcrafting: A 21st-Century Industrial Policy," Roosevelt Institute, May 31, 2023.
18 Hughes and Spiegler, "Marketcrafting."

19 "Investment Needs of USD 35 trillion by 2030 for Successful Energy Transition," International Renewable Energy Agency press release, March 28, 2023.
20 Mark Paul and Nina Eichacker, "Why We Should Be Funding More Solyndras," *MIT Technology Review,* November 19, 2020.
21 Hughes and Spiegler, "Marketcrafting," 36.
22 "Biden-Harris Administration Announces $30 Million to Build Up Domestic Supply Chain for Critical Minerals," Department of Energy press release, August 21, 2023.
23 Katie Meyer, "In a Rare Bipartisan Move, the Pa. Legislature Approves a New Home Repair Assistance Program," WHYY, July 8, 2022.
24 Saule Omarova, "The Climate Case for a National Investment Authority," Data for Progress, August 2020. See also Saule Omarova et al., "The National Investment Authority: A Blueprint," Berggruen Institute, March 4, 2022.
25 Omarova et al., "The National Investment Authority."
26 Olivia Alperstein, "High Flyers 2023: How Ultra-Rich Private Jet Travel Costs the Rest of Us and Burns Up Our Planet," Institute for Policy Studies, May 2023, 10.
27 Alperstein, "High Flyers 2023."
28 Omar Ocampo and Kalena Thomhave, "Private Jets Are Dirty Luxuries for the Ultra Rich. Let's Tax Them," *In These Times,* May 9, 2023.
29 Kate Abnett, "France, Netherlands Call for EU Climate Clampdown on Private Jets," Reuters, May 26, 2023.
30 Harold Meyerson, "Buybacks Are Down, Production Is Up," *The American Prospect,* August 7, 2023.
31 Harold Meyerson, "The Bill That Would Stop Buybacks," *The American Prospect,* May 25, 2023.
32 Lina Khan, "Vision and Priorities for the FTC," Memorandum to the staff and commissioners of the FTC, September 22, 2021.
33 Michael Grunwald, "Inside the War on Coal," *Politico,* May 26, 2015.
34 Seth Feaster, "U.S. on Track to Close Half of Coal Capacity by 2026," Institute for Energy Economics and Financial Analysis, April 3, 2023.
35 Bakatjan Sandalkhan et al., "How Governments Can Solve the EV Charging Dilemma," Boston Consulting Group, October 12, 2021.
36 "Electric Vehicle Charging Stations," US Department of Energy Alternative Fuels Data Center, https://afdc.energy.gov/fuels/electricity_infrastructure.html.
37 Jennifer Conrad, "China Is Racing to Electrify Its Future," *Wired,* June 29, 2022.
38 "Biden–Harris Administration Proposes New Standards for National Electric Vehicle Charging Network," White House press release, June 9, 2022.

39 Bill Visnic, "SAE to Standardize Tesla NACS Charging Connector," *Automotive Engineering,* June 27, 2023.

40 "Interoperability of Public Electric Vehicle Charging Infrastructure," Electric Power Research Institute, August 2019, 7.

41 "Biden-Harris Administration Announces Recipients of Nearly $900 Million for Clean School Buses Under President's Investing in America Agenda," Environmental Protection Agency press release, May 29, 2024.

42 Johanna Bozuwa and Dustin Mulvaney, "A Progressive Take on Permitting Reform: Principles and Policies to Unleash a Faster, More Equitable Green Transition," Roosevelt Institute, August 21, 2023.

43 Bozuwa and Mulvaney, "A Progressive Take on Permitting Reform," 13.

44 "FERC Proposes Interconnection Reforms to Address Queue Backlogs," Federal Energy Regulatory Commission press release, June 16, 2022.

45 "FERC Issues Transmission NOPR Addressing Planning, Cost Allocation," Federal Energy Regulatory Commission press release, April 21, 2022.

46 Emma Penrod, "Why the Energy Transition Broke the U.S. Interconnection System," *Utility Dive,* August 22, 2022.

47 Maine Public Utilities Commission, "Rule Chapters for the Public Utilities Commission," § 65-407, chapter 324 ("Small Generator Interconnection Standards").

48 Victoria Song, "California Passes Right-to-Repair Act Guaranteeing Seven Years of Parts for Your Phone," *The Verge,* September 13, 2023.

49 Emily Heil, "Selling Home-Cooked Food Is Getting Easier, Thanks to Pandemic-Fueled Deregulation," *Washington Post,* October 13, 2021; Mae Anderson, "Farmers Markets Thrive as Customers and Vendors Who Latched On During the Pandemic Remain Loyal," Associated Press, June 26, 2023; "USDA Announces Funding Availability to Expand Meat and Poultry Processing Options for Underserved Producers and Tribal Communities," USDA press release 0085.23, April 19, 2023.

50 M. Ahmed Diomande, James Heintz, and Robert Pollin, "Why U.S. Financial Markets Need a Public Credit Rating Agency," *Economists' Voice,* June 2009.

51 Max Ajl, *A People's Green New Deal* (Pluto Press, 2021), 60–61.

52 Advait Arun, "The Investment Climate," *Phenomenal World,* August 26, 2023.

53 Baysa Naran et al., "Global Landscape of Climate Finance—A Decade of Data: 2011–2020," Climate Policy Initiative, October 27, 2022.

54 Daniela Gabor, "The (European) Derisking State," Center for Open Science, 2023.

55 Skanda Amarnath et al., "Varieties of Derisking," *Phenomenal World,* June 17, 2023.

56 Silvia Amaro, "Europe Shows a United Front Against Biden's Inflation Reduction Act, Says It Threatens Industry," CNBC, November 9, 2022.
57 Christian Davies and Song Jung-a, "South Korea Complains of Growing Friction with US over High-Tech Trade," *Financial Times,* September 18, 2022.
58 Chad P. Bown, "How the United States Solved South Korea's Problems with Electric Vehicle Subsidies Under the Inflation Reduction Act," Peterson Institute for International Economics working paper 23-6, July 2023.
59 Zeyi Yani, "How Did China Come to Dominate the World of Electric Cars?," *MIT Technology Review,* February 21, 2023.
60 Harry Dempsey and Edward White, "China's Battery Plant Rush Raises Fears of Global Squeeze," *Financial Times,* September 4, 2023; Phoebe Sedgman, Jinshan Hong, and Linda Lew, "China's Stranglehold on EV Supply Chain Will Be Tough to Break," Bloomberg, September 27, 2023.
61 Emmet White, "Tesla Exports Cheapest Model Y Yet from China to Canada," *Autoweek,* April 24, 2023.
62 Gavin Bade, "'A Sea Change': Biden Reverses Decades of Chinese Trade Policy," *Politico,* December 26, 2022.
63 Shannon Osaka, "How 'USA-First' Failed the Solar Industry," *Grist,* May 19, 2022; Ben Adler, "Tariffs on Solar Panels Threaten Biden's Climate Change Goals," Yahoo News, May 26, 2022.
64 Van Jackson, *Pacific Power Paradox: American Statecraft and the Fate of the Asian Peace* (Yale University Press, 2023), 180.
65 Ana Swanson and Alan Rappeport, "U.S. Adds Tariffs to Shield Struggling Solar Industry," *New York Times,* June 6, 2024.
66 Charles Sabel and David Victor, *Fixing the Climate: Strategies for an Uncertain World* (Princeton University Press, 2022), 27.
67 Paul J. Young et al., "The Montreal Protocol Protects the Terrestrial Carbon Sink," *Nature* 596 (August 2021): 384–88.
68 Zhenyu Tian et al., "A Ubiquitous Tire Rubber–Derived Chemical Induces Acute Mortality in Coho Salmon," *Science* 371, no. 6525 (December 2020): 185–89.
69 Lisa Winkler et al., "The Effect of Sustainable Mobility Transition Policies on Cumulative Urban Transport Emissions and Energy Demand," *Nature Communications* 14, no. 2357 (April 2023).
70 Eric Larson et al., "Net-Zero America: Potential Pathways, Infrastructure, and Impacts," Princeton University final report summary, October 29, 2021, 20.
71 "Overcoming Inertia in Climate Tech Investing," PwC State of Climate Tech report, November 3, 2022.
72 "Overcoming Inertia in Climate Tech Investing."
73 Daniela Chiriac, Harsha Vishnumolakala, and Paul Rosane, "Landscape of

Climate Finance for Agrifood Systems," Climate Policy Initiative, July 31, 2023.

74 Fredrik Dahlqvist et al., "Climate Investing: Continuing Breakout Growth Through Uncertain Times," McKinsey & Company, March 13, 2023.

75 Brett Christophers, *Our Lives in Their Portfolios: Why Asset Managers Own the World* (Verso Books, 2023), 36.

76 Christophers, *Our Lives in Their Portfolios,* 196.

77 Alexander J. Field, *The Economic Consequences of U.S. Mobilization for the Second World War* (Yale University Press, 2022), 142.

78 Field, *Economic Consequences of U.S. Mobilization,* 95.

79 Eric Williams, *Capitalism and Slavery* (repr. ed., University of North Carolina Press, 1994), 176.

80 Matthew Dalton, "Behind the Rise of U.S. Solar Power, a Mountain of Chinese Coal," *Wall Street Journal,* July 31, 2021.

81 Enrico Mariutti, "The Dirty Secret of the Solar Industry," EnricoMariutti.it, April 7, 2023, 26, https://www.enricomariutti.it/the-dirty-secret-of-solar-industry/.

82 Mariutti, "The Dirty Secret of the Solar Industry," 1.

83 Joseph Rachman, "Indonesia Asks Where the Money Is for Green Transition," *Foreign Policy,* September 26, 2023.

84 International Energy Agency, "CO2 Emissions in 2023," March 2024; see also Emma Penrod, "Investors, Buyers Losing Interest in 'Speculative' Renewable Projects," *Utility Dive,* August 22, 2023.

85 Damian Carrington, "'Insanity': Petrostates Planning Huge Expansion of Fossil Fuels, Says UN Report," *The Guardian,* November 8, 2023.

86 Richard York, "Why Petroleum Did Not Save the Whales," *Socius* 3 (2017): 11.

87 Alejandro Henao and Wesley E. Marshall, "The Impact of Ride-Hailing on Vehicle Miles Traveled," *Transportation* 46 (2019): 2173–94; Alejandro Henao, Wesley E. Marshall, and Bruce Janson, "Impacts of Ridesourcing on VMT, Parking Demand, Transportation Equity, and Travel Behavior," Mountain-Plains Consortium, March 2019; Bruce Schaller, "Can Sharing a Ride Make for Less Traffic? Evidence from Uber and Lyft and Implications for Cities," *Transport Policy* 102 (March 2021): 1–10.

88 John Manuel Barrios, Yael V. Hochberg, and Hanyi Livia Yi, "The Cost of Convenience: Ridesharing and Traffic Fatalities," Chicago Booth research paper 27, March 17, 2019.

89 Timothy Mitchell, *Carbon Democracy: Political Power in the Age of Oil* (Verso Books, 2013), 5 (emphasis in original).

90 Mitchell, *Carbon Democracy,* 80.

91 Rachel Millard, Aime Williams, and Laura Pitel, "Soaring Costs Threaten Offshore Wind Farm Projects," *Financial Times,* August 7, 2023.
92 Alexander Sammon, "Want to Stare into the Republican Soul in 2023?," *Slate,* May 30, 2023.
93 Sammon, "Want to Stare into the Republican Soul in 2023?"
94 Maxine Joselow with Vanessa Montalbano, "Nikki Haley Wants to Address Climate Change Not by Reducing Carbon, But Capturing It," *Washington Post,* February 15, 2023.
95 Sammon, "Want to Stare into the Republican Soul in 2023?"
96 Nick Bowlin, "A Good Prospect: Mining Climate Anxiety for Profit," *The Drift,* July 9, 2023.
97 Bowlin, "A Good Prospect."
98 "A Future Beyond Brick and Mortar—Disruptive Change Ahead in Automotive Retail," McKinsey & Company, September 2020.
99 "Going Small to Go Big: Micromarkets in US Auto Retail and Aftermarket," McKinsey & Company, July 19, 2023.
100 "Going Small to Go Big."

Chapter 2: Public Power

1 Rocío Uría-Martínez and Megan M. Johnson, "U.S. Hydropower Market Report 2023 Edition," US Department of Energy Water Power Technologies Office, 2.
2 University of Michigan Center for Sustainability Studies, U.S. Grid Energy Storage Factsheet.
3 Uría-Martínez and Johnson, "U.S. Hydropower Market Report 2023 Edition," 17.
4 Uría-Martínez and Johnson, "U.S. Hydropower Market Report 2023 Edition," 2.
5 Erik H. Schmidt, Megan M. Johnson, and Rocio Uría-Martínez, "U.S. Pumped Storage Hydropower Development Pipeline Map 2023," HydroSource, Oak Ridge National Laboratory.
6 Katherine Antonio, Jonathan Russo, and Elesia Fasching, "New Pumped-Storage Capacity in China Is Helping to Integrate Growing Wind and Solar Power," U.S. Energy Information Administration, August 9, 2023.
7 Fred Mayes, "Most Pumped Storage Electricity Generators in the U.S. Were Built in the 1970s," U.S. Energy Information Administration, October 31, 2019.
8 "Grid-Scale Storage: Overview," https://www.iea.org/energy-system/electricity/grid-scale-storage#overview.

9 "Abstracts of Papers Presented at the 35th Annual Meeting of the Association of Southeastern Biologists," *ASB Bulletin* 21, no. 2 (April 1974): 88.

10 Daniela Gabor and Benjamin Braun, "Green Macrofinancial Regimes," SocArXiv 4pkv8, Center for Open Science, October 5, 2023, 28.

11 Matthew Huber, *Climate Change as Class War: Building Socialism on a Warming Planet* (Verso Books, 2022), 241.

12 Matt Bruenig, "Fighting Climate Change with a Green Tennessee Valley Authority," People's Policy Project, January 24, 2019.

13 Thea Riofrancos et al., "Achieving Zero Emissions with More Mobility and Less Mining," Climate + Community Project, University of California at Davis, January 2023.

14 Jewellord T. Nem Singh, "Industrial Experiments," *Phenomenal World,* November 30, 2023.

15 Melanie Brusseler and Sandeep Vaheesan, "What If We Just Nationalize the Power Grid?," *Heatmap,* June 6, 2024.

16 Ashik Siddique, "100 Days of Socialism, with the Green New Deal," *Democratic Left,* April 7, 2021.

17 Amy Kapczynski and Joel Michaels, "Administering a Democratic Industrial Policy," Yale Law School public law research paper, January 30, 2024, 42.

18 Amy Kapczynski, Reshma Ramachandran, and Christopher Morten, "How Not to Do Industrial Policy," *Boston Review,* October 2, 2023.

19 Camilla Hodgson, "Wildfires Destroy Almost All Forest Carbon Offsets in 100-Year Reserve, Study Says," *Financial Times,* August 5, 2022.

20 Lisa Song, "An Even More Inconvenient Truth," ProPublica, May 22, 2019; see also Heidi Blake, "The Great Cash-for-Carbon Hustle," *The New Yorker,* October 16, 2023.

21 Ashley Dawson, "How to Win a Green New Deal in Your State," *The Nation,* May 11, 2023.

22 Huber, *Climate Change as Class War,* 219.

23 Alyssa Battistoni, "Living, Not Just Surviving," *Jacobin,* August 15, 2017.

24 Gabriel Winant and Teagan Harris, "Socialists Are Trying to Revive the American Labor Movement," *Jacobin,* May 1, 2022.

25 Todd E. Vachon et al., "Bargaining for Climate Justice," *The Forge,* March 31, 2020.

26 Courtney E. Martin, "Teachers Are Striking for More Than Just Pay Raises," *Vox,* July 16, 2023.

27 Alyssa Battistoni, "Sustaining Life on This Planet," in *Democratize Work: The Case for Reorganizing the Economy,* ed. Isabelle Ferreras, Julie Battilana, and Dominique Méda (University of Chicago Press, 2022), 107.

28 Kim Moody, *Breaking the Impasse: Electoral Politics, Mass Action, and the New Socialist Movement in the United States* (Haymarket Books, 2022), 45.

29 Moody, *Breaking the Impasse,* 171.
30 Cara Buckley, "Facing Budget Shortfalls, These Schools Are Turning to the Sun," *New York Times,* September 15, 2022.
31 Lenore Palladino, "Labor's Green Capital," *Phenomenal World,* September 12, 2023.
32 George R. Stewart, *Storm* (repr. ed., New York Review of Books, 2021), 47.
33 Stewart, *Storm,* 48.
34 Stewart, *Storm,* 190–94.
35 Stewart, *Storm,* 195.
36 Italo Calvino, "The Adventure of a Clerk," in *Difficult Loves* (Picador, 1985), 38.
37 Amelia Urry, "Welcome to Paradise: Batteries Now Included," *Grist,* August 8, 2017.
38 Allan Parachini, "Kauai: World's Biggest Solar Power Plant Relies on a Flock of Sheep," *Honolulu Civil Beat,* January 24, 2019.
39 Brittany Lyte, "The Shift to a Green Energy Future Is Renewing Plantation-Era Water Wars on Kauai," *Honolulu Civil Beat,* March 16, 2023.
40 "Pōʻai Wai Ola/West Kauai Watershed Alliance Intervenes in KIUC Proceeding," *Ililani Media,* January 22, 2021.
41 "Convention (No. 169) Concerning Indigenous and Tribal Peoples in Independent Countries," International Labour Organization, June 27, 1989.
42 Eduardo Medina and April Rubin, "Remains of Nearly 5,000 Native Americans Will Be Returned, U.S. Says," *New York Times,* April 4, 2023.
43 José Carlos Mariátegui, "The Problem of the Indian," in *Selected Works,* ed. and trans. Christian Noakes (Iskra Books, 2021), 37.
44 Ainsley Thomson, "New Zealand's Māori Party Sees Resurgence in Election Result," Bloomberg, October 14, 2023.
45 Thea Riofrancos, *Resource Radicals: From Petro-Nationalism to Post-Extractivism in Ecuador* (Duke University Press, 2020), 49–50.
46 Nimrod Baranovitch, "The Impact of Environmental Pollution on Ethnic Unrest in Xinjiang: A Uyghur Perspective," *Modern China* 45, no. 5 (September 2019): 504–36.
47 Martín Arboleda, *Planetary Mine: Territories of Extraction Under Late Capitalism* (Verso Books, 2020), 221–22.
48 Arboleda, *Planetary Mine,* 223–24.
49 José Carlos Mariátegui, "Nationalism and Vanguardism," in *Selected Works,* ed. and trans. Christian Noakes (Iskra Books, 2021), 103.
50 Anneleen Ophoff, "They Tried to Make Us Tourists in Our Homelands," interview with Áslat Holmberg and Kukka Ranta, *Are We Europe* 15 (August 9, 2022).

51 Katie J. M. Baker and Tom Warren, "WWF Funds Guards Who Have Tortured and Killed People," *BuzzFeed News,* March 4, 2019.
52 Jared Olson, "The Cash Before the Killing," *The Intercept,* June 23, 2022.
53 Saara-Maria Salonen, "Sámi Leaders Voice Concern over Projected Wind Farm," *Barents Observer,* December 1, 2021.
54 Minqi Li, "China: Imperialism or Semi-Periphery?," *Monthly Review* 73, no. 3 (July–August 2021).
55 Minqi Li, *China and the 21st Century Crisis* (Pluto Press, 2016), 74.
56 Li, "China: Imperialism or Semi-Periphery?"
57 Li, *China and the 21st Century Crisis,* 77.
58 Viola Zhou, "TSMC's Debacle in the American Desert," *Rest of World,* April 23, 2024.
59 Melissa Chan and Heriberto Araújo, "China Wants Food. Brazil Pays the Price," *The Atlantic,* February 15, 2020; Guilherme C. Delgado and Sérgio Pereira Leite, "The Agribusiness Pact," *Phenomenal World,* August 5, 2023.
60 Cedric Robinson, *Black Marxism: The Making of the Black Radical Tradition* (repr. ed., University of North Carolina Press, 2000), 42.
61 Robinson, *Black Marxism,* 42.
62 Noel Ignatiev, "Black Worker, White Worker," in *Treason to Whiteness Is Loyalty to Humanity* (Verso Books, 2022), 100.
63 Maria Mies, *Patriarchy and Accumulation on a World Scale: Women in the International Division of Labour* (Zed Books, 1998), 109.
64 Mies, *Patriarchy and Accumulation on a World Scale,* 69.
65 Mies, *Patriarchy and Accumulation on a World Scale,* 109.
66 Stefania Barca, *Workers of the Earth: Labour, Ecology and Reproduction in the Age of Climate Change* (Pluto Press, 2024), 62–76.
67 Ruy Mauro Marini, *The Dialectics of Dependency* (Monthly Review Press, 2022), 130–36.
68 Marini, *The Dialectics of Dependency,* 132.
69 Li, *China and the 21st Century Crisis,* 77.
70 Christine Delphy, "For Equality," in *Separate and Dominate: Feminism and Racism After the War on Terror* (Verso Books, 2015), 51–52.
71 Kai Bosworth, *Pipeline Populism: Grassroots Environmentalism in the Twenty-First Century* (University of Minnesota Press, 2020), 36.
72 Moishe Postone, *Time, Labor, and Social Domination: A Reinterpretation of Marx's Critical Theory* (Cambridge University Press, 1993), 324.
73 "Notes on the New Housing Question," in *Endnotes No. 2: Misery and the Value Form* (April 2010), 66.

74 Hamilton Nolan, *The Hammer: Power, Inequality, and the Soul of Labor* (Hachette Books, 2024), 85.
75 Joshua Clover, *Riot. Strike. Riot: The New Era of Uprisings* (Verso Books, 2019), 147.
76 Clover, *Riot. Strike. Riot,* 147.
77 Judi Bari, *Timber Wars* (Common Courage Press, 1994), 35.
78 Bari, *Timber Wars,* 74.
79 Keith Brower Brown and Sara Holiday Nelson, "Working Sunset to Sunrise: Union Strategies in Three California Climate Transitions," *Environmental Politics* 33, no. 4 (October 15, 2023): 17.
80 Brown and Nelson, "Working Sunset to Sunrise," 8.
81 Brooke Anderson, "Building a Worker-Led Movement for Climate Justice," in *Power Lines: Building a Labor-Climate Justice Movement,* ed. Jeff Ordower and Lindsay Zafir (New Press, 2024), 154.
82 Tobias Kalt, "Agents of Transition or Defenders of the Status Quo? Trade Union Strategies in Green Transitions," *Journal of Industrial Relations* 64, no. 4 (September 2022): 499–521, 516.
83 Kalt, "Agents of Transition or Defenders of the Status Quo?," 517.
84 Paige Oamek, "Electrical Workers Union Fights to Expand Fossil Fuel–Powered Crypto Mining in New York," *New York Focus,* June 14, 2022.
85 Colin Kinniburgh, "Power Industry Quietly Pushes New York to Endorse Non-Renewable Energy," *New York Focus,* March 15, 2022.

Chapter 3: Communism

1 Gracchus Babeuf and the Conspiracy of the Equals, "Manifesto of the Equals," 1796, https://www.marxists.org/history/france/revolution/conspiracy-equals/1796/manifesto.htm.
2 Babeuf and the Conspiracy of the Equals, "Manifesto of the Equals."
3 Carolyn Merchant, *Earthcare: Women and the Environment* (Routledge, 1996), 193.
4 Jasper Bernes, "Between the Devil and the Green New Deal," *Commune,* April 25, 2019.
5 István Mészáros, *Beyond Capital: Toward a Theory of Transition* (Monthly Review Press, 2010), 819.
6 Malcom Ferdinand, *Decolonial Ecology: Thinking from the Caribbean World* (Polity, 2022), 194.
7 M. E. O'Brien, "To Abolish the Family," in *Endnotes No. 5: The Passions and the Interests* (Autumn 2019), 416.

8 David Graeber, "The New Anarchists," *New Left Review* 13 (January–February 2002), 61–73, 61.
9 Graeber, "The New Anarchists," 64.
10 Albert Folch and Jordi Planas, "Cooperation, Fair Trade, and the Development of Organic Coffee Growing in Chiapas (1980–2015)," *Sustainability* 11, no. 2 (January 2019).
11 "San Andrés Accords," Uppsala Conflict Data Program.
12 Carol Hernández, Hugo Perales, and Daniel Jaffee, "'Without Food There Is No Resistance': The Impact of the Zapatista Conflict on Agrobiodiversity and Seed Sovereignty in Chiapas, Mexico," *Geoforum* 128 (January 2022).
13 Hernández, Perales, and Jaffee, "'Without Food There Is No Resistance.'"
14 Shinsuke Ogawa, *Sanrizuka: Heta Village* (Ogawa Productions, 1973), 2 hours, 26 minutes.
15 Ogawa, *Sanrizuka,* 2:01–2.
16 Graeber, "The New Anarchists," 73.
17 Sam Moyo and Paris Yeros, "The Resurgence of Rural Movements Under Neoliberalism," in *Reclaiming the Land: The Resurgence of Rural Movements in Africa, Asia and Latin America* (Zed Books, 2005), 14.
18 Moyo and Yeros, "Resurgence of Rural Movements," 41.
19 Moyo and Yeros, "Resurgence of Rural Movements," 49.
20 Leanne Betasamosake Simpson, *As We Have Always Done: Indigenous Freedom Through Radical Resistance* (University of Minnesota Press, 2017), 76–77.
21 Simpson, *As We Have Always Done,* 79.
22 Simpson, *As We Have Always Done,* 80.
23 Val Napoleon, "Thinking About Indigenous Legal Orders," research paper for the National Centre for First Nations Governance, June 2007, 19.
24 Napoleon, "Thinking About Indigenous Legal Orders," 3.
25 Food and Agriculture Organization of the United Nations, Alliance of Bioversity International, and International Center for Tropical Agriculture, *Indigenous Peoples' Food Systems: Insights on Sustainability and Resilience in the Front Line of Climate Change* (2021), 6.
26 Stephen T. Garnett et al., "A Spatial Overview of the Global Importance of Indigenous Lands for Conservation," *Nature Sustainability* 1, no. 7 (July 2018): 371.
27 Josiane Kouagheu, "Dans le sud-est du Cameroun, les Baka sont marginalisés au nom de la protection de la nature," *Le Monde,* August 20, 2022.
28 Peter Gelderloos, *The Solutions Are Already Here: Strategies for Ecological Revolution from Below* (Pluto Press, 2022), 112–13.

29 Stefania Barca, *Forces of Reproduction: Notes for a Counter-Hegemonic Anthropocene* (Cambridge University Press, 2020), 60.
30 Max Ajl, *A People's Green New Deal* (Pluto Press, 2021), 127.
31 Ajl, *A People's Green New Deal,* 127.
32 Mariama Sonko, "What Kind of Market Do We Need for the Transition to Agro-Ecology?," Alliance for Food Sovereignty in Africa biennial food system conference and celebration, October 2020, via agroecologyfund.org.
33 Patrick Bresnihan and Naomi Millner, *All We Want Is the Earth: Land, Labour and Movements Beyond Environmentalism* (Bristol University Press, 2023), 114.
34 Cira Pascual Marquina and Chris Gilbert, "Venezuela: Food Is Not a Commodity, It's a Human Right: Pueblo a Pueblo Builds Food Sovereignty (Part I)," *Monthly Review,* May 17, 2023.
35 Pascual Marquina and Gilbert, "Venezuela: Food Is Not a Commodity, It's a Human Right."
36 Cira Pascual Marquina and Chris Gilbert, "Growing Native Potatoes in Synergy with the Land and Its People," *Monthly Review,* October 23, 2023.
37 Lynn Corelle and jimmy cooper, eds., *Make the Golf Course a Public Sex Forest* (Maitland Systems Engineering, 2023).
38 Harsha Walia, *Undoing Border Imperialism* (AK Press, 2013), 25.
39 Élisabeth Vallet, "The World Is Witnessing a Rapid Proliferation of Border Walls," Migration Policy Institute, March 2, 2022.
40 Vallet, "The World Is Witnessing a Rapid Proliferation of Border Walls."
41 Joseph Chamie, *Population Levels, Trends, and Differentials: More Important Population Matters* (Springer, 2022), 260.
42 Paul Gilroy, "'Where Every Breeze Speaks of Courage and Liberty': Offshore Humanism and Marine Xenology, or, Racism and the Problem of Critique at Sea Level," lecture delivered at the 2015 Royal Geographical Society–Institute of British Geographers annual international conference, *Antipode* 50, no. 1 (January 2018): 16.
43 Jasmine Aguilera, "Humanitarian Scott Warren Found Not Guilty After Retrial for Helping Migrants at Mexican Border," *Time,* November 21, 2019.
44 Gelderloos, *The Solutions Are Already Here,* 142.
45 Thomas Lacroix, Filippo Furri, and Louise Hombert, "International Migration and the Rise of Urban Militant Networks in the Mediterranean," *Frontiers in Political Science,* August 5, 2022; Sabine Volk, "'No Borders, No Nations' or 'Fortress Europe'? How European Citizens Remake European Borders," in *European Studies and Europe: Twenty Years of Euroculture,* ed. Janny de Jong et al. (Göttingen University Press, 2020), 77–92.

46 Mary Jacobus, "The Perspective of the Other," in "In Dialogue: Perspectives on Migration," Princeton University Press, December 14, 2022.
47 Mariame Kaba, "Yes, We Mean Literally Abolish the Police," *New York Times,* June 12, 2020.
48 Julia Lurie, "They Built a Utopian Sanctuary in a Minneapolis Hotel. Then They Got Evicted," *Mother Jones,* June 12, 2020.
49 Malcolm Harris, "Sweeping Homeless Encampments Is Cruel and Unacceptable," *The Nation,* April 14, 2022.
50 Hilary Klein, *Compañeras: Zapatista Women's Stories* (Seven Stories Press, 2015), 83–88.
51 Bertolt Brecht, "Many are in favor of order . . . ," in *The Collected Poems of Bertolt Brecht* (Liveright, 2018), 471.
52 Abdullah Öcalan, "Self-Protection Is Universal Right," Hawar News Agency, May 17, 2021.
53 Bruno Latour, *Down to Earth: Politics in the New Climatic Regime* (Polity, 2018), 66.
54 Andreas Malm, *How to Blow Up a Pipeline: Learning to Fight in a World on Fire* (Verso Books, 2021), 147.
55 Hilary Beaumont, "The Activists Sabotaging Railways in Solidarity with Indigenous People," *The Guardian,* July 29, 2021; Julia Shipley, "'You Strike a Match': Why Two Women Sacrificed Everything to Stop the Dakota Access Pipeline," *Grist,* May 26, 2021.
56 Anonymous, "Treasure Hunt for Coastal Gaslink," BC Counter Info, February 24, 2023.
57 Andréa Leme da Silva et al., "Water Appropriation on the Agricultural Frontier in Western Bahia and Its Contribution to Streamflow Reduction: Revisiting the Debate in the Brazilian Cerrado," *Water* 13, no. 8 (April 2021).
58 Mike Davis, "Thanatos Triumphant," *Sidecar* (blog of *New Left Review*), March 7, 2022.
59 Davis, "Thanatos Triumphant."
60 Abdullah Öcalan, *Democratic Confederalism* (Transmedia Publishing, 2011), 32.
61 Öcalan, *Democratic Confederalism.*
62 Andreas Chatzidakis, "Commodity Fights in Post-2008 Athens: Zapatistas Coffee, Kropotkinian Drinks and Fascist Rice," *ephemera* 13, no. 2 (May 2013): 459–68.
63 Alex King and Ioanna Manoussaki-Adamopoulou, "Inside Exarcheia: The Self-Governing Community Athens Police Want Rid Of," *The Guardian,* August 26, 2019.
64 Conspiracy of Cells of Fire, "And Death Will No Longer Have Any Power:

Concerning the Fire in the Marfin Bank," July 10, 2010, https://theanarchistlibrary.org/library/conspiracy-of-cells-of-fire-and-death-will-no-longer-have-any-power?v=1717957804.

65 Enrique Dussel, *Twenty Theses on Politics* (Duke University Press, 2008), 105–6.

66 Dussel, *Twenty Theses on Politics,* 105.

67 Timothy Mitchell, *Carbon Democracy: Political Power in the Age of Oil* (Verso Books, 2013), 17.

68 Dilar Dirik, *The Kurdish Women's Movement: History, Theory, Practice* (Pluto Press, 2022), 101.

69 "At Cloverlick: An Interview with the Blackjewel Miners' Blockade," *Pinko* 1 (October 15, 2019).

70 M. E. O'Brien, *Family Abolition: Capitalism and the Communizing of Care* (Pluto Press, 2023), 210.

71 O'Brien, *Family Abolition,* 210.

72 Rosa Luxemburg, "Order Prevails in Berlin," January 14, 1919, https://www.marxists.org/archive/luxemburg/1919/01/14.htm.

73 Ferdinand, *Decolonial Ecology,* 114.

74 Mao Zedong et al., *The Chinese Revolution and the Chinese Communist Party,* 1939, https://www.marxists.org/reference/archive/mao/selected-works/volume-2/mswv2_23.htm; Fidel Castro, "Speech at Moncada Anniversary Ceremony," July 26, 1985, Castro Speech Data Base [sic], Latin American Network Information Center; Amílcar Cabral, *Return to the Source: Selected Texts of Amilcar Cabral,* 2nd new expanded ed. (Monthly Review Press, 2022).

75 Rita Laura Segato, "Gender and Violence in the Apocalyptic Phase of Capital," in *Feminicide and Global Accumulation: Frontline Struggles to Resist the Violence of Patriarchy and Capitalism* (Common Notions, 2021), 64.

76 Vincent Bevins, *The Jakarta Method: Washington's Anticommunist Crusade and the Mass Murder Program That Shaped Our World* (repr. ed., PublicAffairs, 2021), 235–36.

77 Patrícia Godinho Gomes et al., "Causes of Violence Against Women and the Relationship Between the Murder of Women and Global Accumulation," in *Feminicide and Global Accumulation: Frontline Struggles to Resist the Violence of Patriarchy and Capitalism* (Common Notions, 2021), 44.

78 Silvia Federici, "Globalization, the Accumulation of Capital and Violence Against Women: An International and Historical Perspective," in *Feminicide and Global Accumulation: Frontline Struggles to Resist the Violence of Patriarchy and Capitalism* (Common Notions, 2021), 156.

79 Mike Davis, *Old Gods, New Enigmas: Marx's Lost Theory* (repr. ed., Verso Books, 2020), xv.

Chapter 4: The Planetary Crisis

1 See Malcolm Harris, "Just Beans," *The Drift,* November 1, 2022.
2 National Intelligence Council, *Global Trends 2040: A More Contested World,* March 2021, 2.
3 Jeff Goodell, *The Heat Will Kill You First: Life and Death on a Scorched Planet* (Little, Brown, 2023), 311–12.
4 Jonathan Blitzer, "How Climate Change Is Fuelling the U.S. Border Crisis," *The New Yorker,* April 3, 2019.
5 World Bank Group, "Employment in Agriculture (% of Total Employment) (Modeled ILO Estimate)."
6 "Joint Statement from the United States and Guatemala on Migration," White House Briefing Room, June 1, 2023.
7 National Intelligence Council, *Global Trends 2040,* 101.
8 Blitzer, "How Climate Change Is Fuelling the U.S. Border Crisis."
9 International Organization for Migration, "US-Mexico Border World's Deadliest Migration Land Route," September 12, 2023.
10 Forensic Architecture, "Drift-Backs in the Aegean Sea," July 15, 2022.
11 Forensic Architecture, "Drift-Backs in the Aegean Sea."
12 Nicola Kelly, "Revealed: UK-Funded French Forces Putting Migrants' Lives at Risk with Small-Boat Tactics," *The Guardian,* March 23, 2024.
13 Tom Philpott, *Perilous Bounty: The Looming Collapse of American Farming and How We Can Prevent It* (Bloomsbury, 2020), 187.
14 Zaid Ali Basha, "The Agrarian Question in Yemen: The National Imperative of Reclaiming and Revalorizing Indigenous Agroecological Food Production," *Journal of Peasant Studies* 50, no. 3 (2023): 879–930.
15 Daniel Grossman, "Water War: Is Big Agriculture Killing Brazil's Traditional Farms?," *Yale Environment 360,* November 10, 2021.
16 Juliana Siqueira-Gay and Luis E. Sánchez, "The Outbreak of Illegal Gold Mining in the Brazilian Amazon Boosts Deforestation," *Regional Environmental Change* 21 (March 17, 2021).
17 Clarissa Gandour et al., "Forest Degradation in the Brazilian Amazon: Public Policy Must Target Phenomenon Related to Deforestation," Climate Policy Initiative, March 9, 2021.
18 Dom Phillips et al., "Revealed: Rampant Deforestation of Amazon Driven by Global Greed for Meat," *The Guardian,* July 2, 2019.
19 Marion Renault, "Animals Are Dying in Droves. What Are They Telling Us?," *New Republic,* May 3, 2023.
20 Sophie Chao, *In the Shadow of the Palms: More-Than-Human Becomings in West Papua* (Duke University Press, 2022), 103.

21 Chao, *In the Shadow of the Palms,* 103.
22 Irmak Ertör, "'We Are the Oceans, We Are the People!': Fisher People's Struggles for Blue Justice," *Journal of Peasant Studies* 50, no. 3 (December 1, 2021): 1157–86.
23 Zhibo Luo, Shanying Hu, and Dingjiang Chen, "The Trends of Aquacultural Nitrogen Budget and Its Environmental Implications in China," *Scientific Reports* 8 (July 18, 2018).
24 Ian Urbina, "The Crimes Behind the Seafood You Eat," *The New Yorker,* October 9, 2023.
25 Esther Htusan and Margie Mason, "More Than 2,000 Enslaved Fishermen Rescued in 6 Months," Associated Press, September 17, 2015.
26 Sarah M. Glaser, Paige M. Roberts, and Kaija J. Hurlburt, "Foreign Illegal, Unreported, and Unregulated Fishing in Somali Waters Perpetuates Conflict," *Frontiers in Marine Science* 6 (December 2019).
27 Abdirahman J. Kulmiye, *Untapped Potential Held Back by Skills Shortage* (Heritage Institute for Policy Studies and City University of Mogadishu, May 2020).
28 Charles Kilgour and Duncan Copeland, "Illegal Fishing Hotspot Identified in Northwest Indian Ocean," Global Fishing Watch and Trygg Mat Tracking, June 29, 2020.
29 Kulmiye, *Untapped Potential Held Back by Skills Shortage.*
30 Mark Duffield and Nicholas Stockton, "How Capitalism Is Destroying the Horn of Africa: Sheep and the Crises in Somalia and Sudan," *Review of African Political Economy* (November 1, 2023).
31 Shae Frydenlund and Elizabeth Cullen Dunn, "Refugees and Racial Capitalism: Meatpacking and the Primitive Accumulation of Labor," *Political Geography* 95 (May 2022).
32 Chico Harlan, "For Somalis, Hope Falls to the Cutting Floor," *Washington Post,* May 24, 2016.
33 Alex Blanchette, *Porkopolis: American Animality, Standardized Life, and the Factory Farm* (Duke University Press, 2020), 189.
34 Blanchette, *Porkopolis,* 161.
35 Naomi Klein, *The Shock Doctrine: The Rise of Disaster Capitalism* (Metropolitan Books, 2007), 9.
36 Anthony DiMauro, "The Disaster Vultures," *Business Insider,* August 31, 2023.
37 Emily Atkin, "Climate Change and the Death of the Small Farm," *New Republic,* March 27, 2019.
38 Vandana Shiva, *Earth Democracy: Justice, Sustainability, and Peace,* 2nd ed. (Zed Books, 2016), 16.
39 Erick Burgueño Salas, "Economic Losses from Natural Disaster Events Worldwide from 2000 to 2023," Statista, February 21, 2024.

40 "Climate Change Is Coming for America's Property Market," *The Economist,* September 21, 2023.

41 Michael R. Blood, "California Insurance Market Rattled by Withdrawal of Major Companies," Associated Press, June 5, 2023.

42 Maximilian Kotz, Anders Levermann, and Leonie Wenz, "The Economic Commitment of Climate Change," *Nature* 628 (April 17, 2024): 551–52.

43 Stephen Morgan, Keith Fuglie, and Jeremy Jelliffe, "World Agricultural Output Growth Continues to Slow, Reaching Lowest Rate in Six Decades," USDA Economic Research Service, December 5, 2022.

44 Ziming Zhou, Zhiming Yu, and Sihan Gao, "Climate Shocks and Farmers' Agricultural Productive Investment: Resisting Risk or Escaping Production?," *Frontiers in Ecology and Evolution* 10 (April 2022).

45 "Global Economy Set for Weakest Half-Decade Performance in 30 Years," World Bank Group press release, January 9, 2024.

46 Jennifer Sherer and Nina Mast, "Child Labor Laws Are Under Attack in States Across the Country," Economic Policy Institute, March 14, 2023.

47 "50 Million People Worldwide in Modern Slavery," International Labour Organization press release, September 12, 2022.

48 Aleks Phillips, "Thousands of Solar Panels in Texas Destroyed by Hailstorm," *Newsweek,* March 26, 2024.

49 Christian Parenti, *Tropic of Chaos: Climate Change and the New Geography of Violence* (Nation Books, 2011), 11.

50 Andrew Stroehlein, "The EU Rewards Repression," Human Rights Watch Daily Brief, March 18, 2024.

51 Andrei Popoviciu and José Bautista, "How an EU-Funded Security Force Helped Senegal Crush Democracy Protests," Al Jazeera, February 29, 2024.

52 Ali Kadri, *The Accumulation of Waste: A Political Economy of Systemic Destruction* (Brill, 2023), 21.

53 Kadri, *The Accumulation of Waste,* 23.

54 Mike Stone, "Wars Raise Profit Outlook for US Defense Industry in 2024," Reuters, December 18, 2023.

55 Stephen Losey, "Businesses Reposition amid Growing Demand for Solid Rocket Motors," *Defense News,* October 31, 2023.

56 Losey, "Businesses Reposition amid Growing Demand."

57 William J. Nuttall, Constantine Samaras, and Morgan Bazilian, "Energy and the Military: Convergence of Security, Economic, and Environmental Decision-Making," working paper EPRG 1717, Energy Policy Research Group, Cambridge Judge Business School, University of Cambridge, November 2017;

Neta Crawford, *The Pentagon, Climate Change, and War: Charting the Rise and Fall of U.S. Military Emissions* (MIT Press, 2022), 6.

58 Zetkin Collective and Andres Malm, *White Skin, Black Fuel: On the Danger of Fossil Fascism* (Verso Books, 2021), 476. In the quoted passage, the authors are actually evaluating the likelihood of *ecological* fascism, but I believe one would use the same tools for the fossil variety. I beg the reader's pardon for my manipulation, which is in the interest of concision.

Conclusion

1 Lorraine Hansberry, adapted by Robert Nemiroff, *To Be Young, Gifted and Black: Lorraine Hansberry in Her Own Words* (Vintage Books, 1995), 212–13.

2 Philip Ball, "Is Photosynthesis Quantum-ish?," *Physics World,* April 2018.

3 István Mészáros, *Beyond Capital: Toward a Theory of Transition* (Monthly Review Press, 2010), 908.

4 Lan Xiaohuan, *How China Works: An Introduction to China's State-Led Economic Development* (Palgrave Macmillan, 2024), 154.

5 Lan Xiaohuan, *How China Works,* 118.

6 Wei Zhang, "China's Health and Health Care in the 'New Era,'" *Monthly Review,* October 1, 2023.

7 Thandika Mkandawire, "Thinking About Developmental States in Africa," *Cambridge Journal of Economics* 25, no. 3 (May 2001): 289–314.

8 Daniela Gabor and Ndongo Samba Sylla, "Derisking Developmentalism: A Tale of Green Hydrogen," *Development and Change* 54, no. 5 (September 2023): 1169–96, 1192.

9 Jonathan Weisman, "Flush with Federal Money, Strings Attached, a Deep South Factory Votes to Unionize," *New York Times,* May 12, 2023.

10 Simon Romero, "Brazil Gripped by General Strike over Austerity Measures," *New York Times,* April 28, 2017.

11 André Singer, "Lula's Return," *New Left Review* 139 (January–February 2023), 5–32, 8, 10.

12 Jon Lee Anderson, "The Brazilian Special Forces Unit Fighting to Save the Amazon," *The New Yorker,* April 1, 2024.

13 Jeff Engels, "Twenty Years Later, Remembering the Battle in Seattle," *Labor Notes,* November 22, 2019.

14 "How Do You Form a Workers Militia?," *La Verité,* August 4, 1944, https://www.marxists.org/history/france/resistance/trotskyists/workers-militia.htm.

15 Suh Jae-jung, "Behind Bellicose Bluster, N. Korea Is Turning Airfields into Greenhouse Farms," *Hankyoreh,* February 12, 2024.

16 New Era Windows Cooperative, "Our Story," NewEraWindows.com.

17 Francesca Gabbriellini and Giacomo Gabbuti, "Italy's Longest-Ever Factory Occupation Shows How Workers Can Transform Production," *Jacobin,* April 4, 2023.
18 "100 × 10,000 – #we Rise," #insorgiamo, accessed October 6, 2024, https://insorgiamo.org/100x10-000/.
19 Frank Egan, "Chestnut Trees Thrive at Harrison Power Station," WDTV, December 14, 2023.
20 Dean Spade, *Mutual Aid: Building Solidarity During This Crisis (and the Next)* (Verso Books, 2020).
21 Brian Merchant, "Torching the Google Car: Why the Growing Revolt Against Big Tech Just Escalated," *Blood in the Machine,* February 14, 2024.
22 Malcolm Harris, "Antifascists Have Become the Most Reasonable People in America," *Pacific Standard,* February 7, 2017.
23 Grace Glass and Sasha Tycko, "Not One Tree," *n+1,* Fall 2023.
24 Malcolm Harris, "Just Beans," *The Drift,* November 1, 2022.
25 Chris Gilbert, "Luisa Cáceres: Commune-Building in Urban Venezuela," *Monthly Review,* December 1, 2022.
26 Tess McNamara, "Crisis of Urban Agriculture: Case Studies in Cuba," *Tropical Resources* 36 (2017).
27 Sinan Koont, *Sustainable Urban Agriculture in Cuba* (University Press of Florida, 2011), 86.
28 "A New Addition to Martin Luther King's Legacy," NPR, January 15, 2007.
29 Walter Benjamin, "Theses on the Philosophy of History," in *Illuminations: Essays and Reflections* (Schocken, 2007), 257–58. For a reapplication of Benjamin's angel to the present moment, see Ajay Singh Chaudhary, *The Exhausted of the Earth: Politics in a Burning World* (Repeater, 2024), 246–48.

INDEX

Note: Italic page numbers refer to illustrations.

Abad, Pacita, 108n
Abe, Shinzo, 172
abolition of slavery, 72–73, 187, 252
accounting practices, 35
administrative control, continuum of, 98
Aegean Sea, 207
affirmation trap, public power and, 130–36
Afghanistan, 165, 220
AFL-CIO, 135
Africa
  migration in, 230
  negative agricultural efficiency and, 225
African National Congress, 170
agribusiness projects, 215–16
agricultural products
  climate change and, 206
  indirect consumption of fossil fuels and, 68
  migrants and, 215
agriculture
  extractive practices and, 209–10, 212–13
  knowledge-intensive agriculture, 162
  natural disasters and, 225–26
  subordinated by capitalism, 150, 153, 155, 160
  Value and, 209, 210, 212, 215–16
agrifood systems, 70
agrobiodiversity management, 149–50
agroecology
  communism and, 149–50, 153–62, 168, 173
  proposals of, 234n
  public power and, 257
  techniques of, 216
Agua Zarca dam, Honduras, 120
aircraft sales, taxes on, 49
airports, private flights supported by public goods, 49–50
air travel
  jet fuel as specialty product, 40, 48–49
  marketcraft and, 48–50
  private flights, 48–50
Ajl, Max, 61–62, 157–58
Al-Bachar, Abderrahim, 20–21
Alice's Farm, 31–33, 107
Alipay app, 54
Al Jazeera, 229–30
"All cops are bastards" (ACAB) slogan, 167
Allende, Salvador, 188
alt-right, 254
Altvater, Elmar, 9, 10n
Amazon (firm), 50, 105, 131
Amazon (geographical area)
  agriculture and, 212
  deforestation in, 126, 155, 212
  selective logging of, 212
American Clean Power Association, 34
American Friends Service Committee, 255
Amini, Mahsa, 174
anarchism
  alt-right and, 254
  anarchist action and, 178, 249
  arson and, 177, 178
  on collective problems, 26, 27
  migrants and, 165, 176
  new anarchists, 147–53
  Stop Cop City in Atlanta and, 255
  violence and, 173
Anderson, Laurie, 18
Anduril, 231
animals
  global industrial animal production, 219–22
  human reaction to behavior of, 213–16
  humans as animals, 213, 214, 216, 222
  mass mortality events and, 214–15
  oil spills and, 39, 45
  planetary crisis and, 213–22, 213n

INDEX

anticonformists, 18
antienvironmental consumer backlash, 41
antifeminist reactionaries, 241
Anti Raids Network, 165
antitrust enforcement, 50–51, 80, 253
Aotearoa, New Zealand, 118
Apoism, 174, 181
aquaculture, 216
Árbenz, Jacobo, 188
Arboleda, Martín, 119
Argentina, 147
Aronoff, Kate, 90n
arson, 171, 177, 178
Asian economies, competition with, 22
asset funds
  capital and, 92
  climate-related investments and, 70, 73
Association Nationale des Villes et Territoires Accueillants, 165
As You Sow, 6
Athens, Greece
  arson in, 177
  communist-aligned demonstrators in, 186
  Exarcheia as insurgent zone, 176, 178
  squatting in, 165, 176
Atlanta, Georgia, 255
atmospheric carbon, 7, 9, 9n, 10, 40, 101. *See also* decarbonization
Australia, 65
authoritarianism, 180, 199–200, 229–30
automobiles. *See also* electric vehicles (EV)
  aftermarket value of, 80
  autoworkers and, 104
  car-sharing platforms and, 75–76
  decarbonization replacement plan and, 68–69, 74
  marketcraft transition role of, 70–71
  National Automobile Dealers Association and, 80–82
  postwar manufacturing of, 71
  public-power planning and, 96–97
  public subsidization of, 68
  reduction in use of, 68–69
  World War II restrictions on travel and, 71–72

Babeuf, Gracchus, 185
Bahia, Brazil, 156, 171
Baka peoples, 155–56
banks
  deregulation of investment banks, 39
  green banks, 36, 44, 45–46
  private efforts financed by, 44
Barca, Stefania, 128–29, 157
Bari, Judi, 133–34, 188n
Battistoni, Alyssa, 104–05, 106
Belt and Road Initiative, 66
Benanav, Aaron, 22
Benjamin, Walter, 258–59
Berkman, Alexander, 172
Berman, Elizabeth Popp, 11n
Bernes, Jasper, 142–43
Bevins, Vincent, 188–89
Beyond Coal initiative, 51, 80
Biden, Joe
  climate-critical minerals and, 43, 64
  critical-minerals free trade agreements and, 64
  decarbonization and, 53
  electric vehicles and, 54
  left-liberal politics on climate crisis, 25
  marketcraft and, 40, 43
  on migration, 206
  1 percent tax on buybacks and, 50
  spending packages of, 38
  tariffs and, 66
big green state plan, 95
Bipartisan Infrastructure Law, 54
birds, 39, 45, 213–14
Black Mesa Pumped Storage Project, 115n
Black Panthers, 186
BlackRock, 5
Blanchette, Alex, 220–22
Blitzer, Jonathan, 205–06
Bloomberg, Michael, 51, 80
bokashi, 159–60
Bolshevik Revolution, 175n, 187
Bolsonaro, Jair, 172
border imperialism, 63, 163–65, 166, 167, 229–30, 237, 241
Boston Consulting Group, 53
Bosworth, Kai, 130
bourgeois revolution, 180
Bowlin, Nick, 81
Braun, Benjamin, 95
Brazil
  abolition in, 187
  Correntina water war in, 171
  food system and, 212
  Indigenous self-management in, 156–57
  trade unions in, 246–47
  violence against women in, 190n
Brazilian Landless Workers' Movement, 147, 156
Brecht, Bertolt, 168
Bresnihan, Patrick, 159–60
British Industrial Revolution, 75
Brose, Chris, 231
Brown, John, 252
Brown, Roger, 25
Bruenig, Matt, 95
Build Public Renewables Act (BPRA), 102, 135
Burkina Faso, 188
Buy European Act, 63

Cabral, Amílcar, 188
Cáceres, Berta, 120
California, 210–11, 212, 224
Calvino, Italo, 112

INDEX

Cambodia, 66
Cameroon, 120
Canada, 65, 161
Canary Islands, 164
capital
  advancement of, 155–56
  asset funds and, 92
  care weaponized by, 150
  cheap-chasing capital, 99
  counterforce to, 83
  death capital, 232
  demands of, 68, 82
  domestic capital, 61
  environmental movement aligned with, 133
  exploitation of cheap nature, 74, 77, 98, 131, 142, 143, 148, 151, 210–11, 212, 214–15, 217, 224, 226
  fish and seafood production and, 218
  food system divorced from land and, 209, 210–11, 212
  fossil capital, 9–10, 9n, 232
  fossil fuel investments and, 105, 246, 251
  green capital, 246
  inhuman dictates of, 93
  labor power and, 134, 175, 230
  life choices controlled by, 101
  market for food and, 143
  meat and poultry industry and, 221
  mobile capital, 63–64
  monopolization and, 50
  natural disasters and capital accumulation, 222–27, 231
  personnel of, 171
  private capital fulfilling public wants and needs, 38
  profit rates of, 237
  public power as opposite of, 102–07
  responsibility for accidents, 178
  scarcity of, 230
  transition and nationalism, 228, 229
  transnational capital, 63, 64
  US government's relations with, 61
  venture capital, 69
  withholding of, 79, 82
  zero-sum situations and, 252
capital flight, 79, 114
capitalism
  accidental deaths caused by, 177
  agriculture subordinated by, 150, 153, 155, 160
  appropriating capitalist commodities, 167
  Cold War and, 189
  communes within, 182–83, 256
  communism and, 141, 142, 163, 168–69, 181, 190
  conditional access to, 17–19, 25
  core/center and periphery of, 124, 125, 164–65, 180
  derisking strategy and, 63–64
  as ecocide pact, 169
  expansion of, 14–15
  externalized costs and, 11
  as form of social metabolism, 14–15, 17, 18–19, 24, 94, 125, 131, 141, 143, 175, 180, 190, 215, 239, 261
  global capitalist exploitation, 125
  as global problem, 14, 26, 61, 66, 67
  history of, 18, 102–03
  industrial capitalism, 126
  inequality maintained within, 16, 21, 122, 126, 141, 163
  innovations adopted by, 33
  international hierarchy of, 122, 124–25, 126n, 131
  isolation of production and consumption, 202
  labor power's bargaining within, 130, 132
  lack of structural challenges to, 18
  law and, 158
  the left and, 61, 251–52
  oil industry and, 4, 5, 6–7
  ownership of large public companies and, 6
  planetary character of, 15n
  police role in, 166–68
  postwar car-centric capitalism, 70–71
  poverty reduction and, 21
  as production for its own sake, 10
  progressive liberals and, 14, 24, 232
  property claims of, 169
  racial access to, 126–27
  resistance to, 151
  sociopolitical change and, 201–02
  sole proprietorships and, 9n
  spaces of rupture in, 176
  technocrats and, 103
  technological innovation harnessed by, 32
  Value and, 116, 123, 133, 139–40, 143, 144, 169, 177, 187, 216, 253
  Value-Life link and, 15, 16
  working-class people and, 16, 106, 121, 122, 126–27, 148
capitalists
  authorities as servants of petrocapitalists, 7
  British slave industry and, 72–73
  buy-in of, 41
  class betrayal and, 251
  collective resources deployed by, 36
  collusion of, 80
  communists and, 251, 253
  coordination of electricity market and, 58
  democracy and, 77–79, 82
  dispossession of, 102, 163
  environmental philosophy and, 23
  freedom defined by, 101
  free trade and, 158
  global production for Western consumers, 123–24
  Indigenous capitalists, 245

INDEX

capitalists *(Cont.)*
  labor power and, 9, 9n, 10, 77–78, 103–04, 106, 121, 122, 130, 217, 230
  marketcraft and, 48, 76–83
  market incentives and, 75
  means of production owned by, 9, 16, 76–77, 98, 102
  measurement of costs and, 10–11
  motivations of, 36–37
  offloading costs and risks to labor, 227
  oil industry and, 78
  political challenges and, 81
  profitability standards of, 44, 45, 59, 76, 82, 94, 98–99
  protocapitalists and, 187
  public power and, 94–95, 96, 97, 99, 100, 102, 246
  pumped-storage hydropower and, 92–93
  representative institutions dominated by, 103
  role in directing society, 59
  social planning and, 97, 102
  social resources controlled by, 78–80, 82, 83, 93, 94
  subsidies and, 40, 42, 44–45, 47, 82, 99, 103, 226
  tax on buybacks and, 50
capital strike, 79
Captain Planet cartoons, 197
carbon capture, 69
carbon democracy, 78, 199, 228, 232
carbon dioxide levels, 62
carbon neutrality, 33
carbon-pricing approach, 36
carbon storage industry, 59
carbon tax proposals, 11, 234n
care. *See also* health-care system
  weaponization of, 150
care and service workers, high-touch care workers, 100, 104–06
CARES Act, 38
Cargill, 253
cassowaries, 215–16
Castro, Fidel, 188
CATL, 65
Central America
  de-development and, 220
  escaped slave communities in, 147
  US immigration policy in, 206
Central American Bank for Economic Integration, 120
Chad, 19–24, 102, 139
CHAdeMO, 54
Chamie, Joseph, 164
Chao, Sophie, 215–16
Chartist movement, 185
Chatzidakis, Andreas, 176
Chevron, 6, 39
Chiapas, Mexico, Zapatistas of, 148–50, 156, 168, 174n, 176, 181, 181n
childcare, 100, 131, 165
child labor, 226–27
Chile, 97, 188, 224
China
  agricultural productivity and, 225–26
  anode makers of, 104
  clean tech of, 63
  coal boom in, 125
  coal-fired electricity in, 73
  decarbonization tools and, 243–44
  electric vehicles and, 53, 54, 64–65, 65n, 66
  Fengning pumped-storage power station, Hebei Province, 90
  global capital of, 66
  health care in, 244–45
  labor's terms of trade (LToT) and, 124–25
  marketcraft and public-power interaction in, 243–44
  marketcraft policies of, 38, 65
  ocean extraction and, 216, 217
  pumped-storage hydropower and, 89–90, 92
  solar panels and, 37, 66, 73, 243
  strategic economic partners and, 66
  unequal US exchange relationship with, 129
  US export restrictions on high-tech materials and, 66
  Xinjiang Uyghur Autonomous Region and, 119
China National Petroleum Corporation, 21
Chinese Communist Party (CCP), 65
CHIPS and Science Act, 40, 125
Chobani, "Dear Alice" video, 32
Christophers, Brett, 70, 92
cigarettes, 68
Cilento, Mario, 135–36
Ciudades Refugio, 165
civil disobedience, 170
class position. *See also* working-class people
  balance of class power and, 241–42
  capitalist class betrayal, 251
  class compromise and, 103
  class conflict and, 80, 103, 185–86, 227, 228, 230, 246
  climate change and, 13
  Democratic Party as capitalist class-conciliationist tool, 106
  divisions within, 126–28
  gender divides in, 127–29
  improving position and, 131
  inclusion through exclusion and, 132n
  international division of labor, 77–78
  labor-capital class dichotomy and, 130
  labor power and, 9, 10, 16, 78, 105, 106, 126
  leverage against ownership class, 83, 105
  natural disasters and capital accumulation, 223, 231

private flights and, 50
public power and, 102, 103, 105–06
ruling class competition and, 80
un/equal exchange between classes and, 9, 9n, 16, 76, 77–78, 99, 103, 122–23, 131–32
clean energy
food system and, 209
infrastructure for, 227
investment in, 246
long-term investment in, 36, 42
permitting reform and, 56
clean-power projects, 33, 34
clean tech, global market share in, 63
climate adaptation, 48, 62–63
Climate and Community Project (now Climate & Community Institute), 56, 57, 96
climate authoritarianism, 199
climate catastrophe, avoidance of, 33
climate change. *See also* global warming
agriculture and, 225–26
atmospheric carbon levels and, 7
building resilience to, 62
class position and, 13
collective responses to, 27–28
declining rainfall and, 212
destabilization of climate, 209
as global problem, 7, 26, 61, 228–29
human changes and, 213n
inadequate responses to, 7, 15–16
lack of self-evident solution to, 12
land-food system's vulnerability to, 213
mediation of, 237
migration and, 63, 164, 200, 204–08
as science, 41
technological innovation in fight against, 99
unpredictability of, 109–10
world economy reduction and, 224–25
climate crisis
approaches to, 24–28
carbon-pricing approach and, 36
cause of, 13, 24
communism and, 168–69
demand-side industrial policy and, 43
democratic approach to, 68
food system and, 212
freedom connected to, 101
Indigenous peoples and, 117–18
international efforts and, 228–29
István Mészáros on, 14
private capital and, 38
public power and, 94
reality of, 16, 25
relevant responses to, 12
societal defense from, 169
subsidies and, 41, 48
supply-side industrial policy and, 36, 43
universal terrestrial progress and, 25
climate discourse, ambiguity within, 61
climate finance, global flows of, 62
climate movement, tactics of, 169–71
climate policy
asylum norm as form of, 208
capitalism and, 61–62, 70
marketcraft and, 63–64
Climate Policy Initiative, 62, 70
climate protests, 167, 199, 199n
climate refugees, 63, 205, 207–08
climate reparations, 237
climate science, 16
climate-strike days, 170
climate targets, 97
Clover, Joshua, 132–33
Cloverlick mine, Harlan County, Kentucky, 182
coal consumption, 75
coal emissions, 51–52
coal industry, 78
coalitions, 27, 235–36, 239, 245, 255
Coastal GasLink Pipeline, 170–71
cobalt, 42
coffee
Nicaraguan coffee, 256
North American Free Trade Agreement and, 148
Salvadoran coffee, 247–48, 256
coherence of partisans
disaster councils and, 258–62
metastrategy of, 237, 253–58
points of, 240–43, 247
problem solving and, 234–35, 234n
quantum walk and, 236–38, 260
Cold War, 148, 164, 186, 188, 197, 256
collective problems, approaches to, 24–25, 26, 27–28
Colombia, 149n, 189–90, 190n
colonialism
communal ways of living attacked by, 187
distribution of power and resources, 241
environmental fragility and, 62
neocolonial structural adjustment, 176–77
public power and, 114–15, 116, 121
slavers of West Indies and, 72–73
World War II and, 188
Colorado River, 211
Combined Charging System (CCS), 54
common good, 140–41
common land, 141
communism
accidents and, 175–79
agroecology and, 149–50, 153–62, 168, 173
"All cops are bastards" (ACAB) slogan, 167
alternative models of communal life, 151
alt-right and, 254
anticommunist purges and, 184–91
capitalism and, 141, 142, 163, 168–69, 181, 190
capitalists and, 251, 253

communism *(Cont.)*
climate crisis and, 168–69
on climate migration, 164
the commune and, 139–46, 141n, 147, 160, 181–84, 231, 256–57
complicity and, 239–40
context of, 191
emergency scenarios and, 254, 262
equality of all people and, 180, 182
family abolition and, 145–46
globalization movement and, 147–48
history of term, 140, 140n
implications of, 142
Indigenous forms of communal resistance, 119, 147–48, 150, 151, 152
legitimacy of, 169, 179–84
localism and, 145
marketcraft and, 251–53, 256–58
Karl Marx's definition of, 158n
moral humanism and, 171
on needs, 143–44
new anarchists and old communists, 147–53
orienting principle of, 143
planning and, 152
police and, 166–68, 240
politics and, 176–77
public power and, 140, 246–51
rose theory and, 168–74, 179, 180
sacrifices and, 183–84
shock doctrine and, 223
social metabolism and, 140, 141, 142, 144, 145, 152, 155, 160–62, 163, 175, 181, 249, 258
social planning and, 152, 162, 171, 177–78
strategic action and, 234
taboo against targeting human beings for violence, 171–72
transition to, 162–68
verve and, 181–82, 255
Communist Party of El Salvador, 187
community disaster councils, 261–62
complicity, fears of, 239–40
Confederation of Indigenous Nationalities of Ecuador, 118
Congo, Democratic Republic of, 155–56, 188, 220
conservation philanthropy, 156
conservatives, liberals joining with, 32–33
Conspiracy of Cells of Fire, 177
Conspiracy of the Equals, 180, 185
consumer-protection violations, 51
consumers
antienvironmental backlash, 41
consumer politics, 256
energy-transition subsidies for, 43
global production for Western consumers, 123–24
tax rebates for, 43
consumption
clean-hands conscious consumerism, 248
freezing consumption of fossil fuels, 62
global distribution of, 61–62
household consumption of fossil fuels, 13, 68
individual conscious consumerism, 252
production separated from, 13, 14
of working-class people, 124
zero-waste consumption, 234n
Cooper's Hill, Gloucester, England, 87–88
Coordinadora Nacional de Trabajadores de la Educación, 248
copper, 42, 119, 161, 225
corporate consolidation, 50, 80
corporate minimum tax, 38
Correntina water war, 171
Costa Rica, 159
COVID-19 pandemic
care workers and, 105
cottage food producers and, 59
George Floyd demonstrations and, 166
marketcraft and, 38
market fundamentalism and, 37
as mass mortality event, 214
public-power perspective on, 98
vaccine development and, 98–99
cryptocurrency mines, 135
Cuba, 200, 220, 257–58
cultural values, 198–99
Cusack, John, 230–31

Dakota Access Pipeline, 118, 170–71
dark marketcraft, 48–49
Data for Progress, 45
Dávila, Ana, 160
Davis, Mike, 27, 171–72, 190
decarbonization
automobile use and, 68–69, 74
autoworkers and, 104
in China, 243–44
communism and, 142
economic planning for, 95
of electricity-generation system, 40, 41, 53, 58, 59, 68
failure of market mechanisms for, 95
funding of, 62
as global issue, 61
in Malawi, 245
marketcraft strategies for, 40–43, 66, 238
planetary crisis and, 196
proposals for, 234n
public investments in, 45–47, 47n, 48, 53, 58, 63, 94
public-power perspective on, 100, 101
replacement plan and, 74
as shared need, 48
social metabolism and, 74, 201

INDEX

support for, 240
of transportation, 41, 53, 58
Declaration of the Rights of Man and Citizen, 140
de-development, 218, 219–20, 241
Deepwater Horizon oil spill, 39
Deese, Brian, 33–34
Deligiorgis, Antonis, 164–65
Delphy, Christine, 129
demand-side industrial policy, 43
Démé, Souleymane, 19
democracy
capitalists and, 77–79, 82
carbon democracy, 78, 199, 228, 232
as oil, 78
planetary crisis and, 199–200
production and, 40
public power and, 94–95, 102, 103, 121–22, 144
representative democracy, 180
responsibility for outcomes and, 58
rules written by, 36, 50, 76
values of, 229
democratic confederalism, 173–74
Democratic Party, 37, 38, 106
Democratic Socialists of America (DSA), 97, 101–02, 105, 135
denial-and-delay complex, 198
Dial, Thornton, *Refugees Trying to Get to the United States* painting, 165–66
Diesel, Vin, 183
DiMauro, Anthony, 223
direct action theory, 170
Dirik, Dilar, 181
disaster councils, 258–62
disorder, fears of, 239, 253, 253n, 254–55
diversity of tactics, 235–36
Dochuk, Darren, 8n
double participation ladder models, 160–61
drift-backs, 207
Driver, Adam, 108
drug pricing reform, 38
Drugs-Value-Life chain, 19
Dry Ground Burning (film), 182–83
Dubai, United Arab Emirates, 219, 219–20n
Duffield, Mark, 219
Durbin, Jon, 250–51
Dussel, Enrique, 178

earthcare ethic, 142, 145, 155, 157–58, 169, 180
Earth First!, 133–34, 188n
Eastern Europe, 165
East Timor massacre, 188
ecological sustainability, 45, 149, 154
economic actors, interchange of, 35
economic planning, 95
Economic Policy Institute, 226
economic power, 17
economic recovery, 37
Economic Research Service, 225
economics, 12
economic shock therapy, 222–27
Ecosocialist Working Group, 97, 101–02
ecosocial metabolism, 150, 160–62, 163, 200, 208–09, 228, 249
ecosystems, 214
education
public education system, 52
public-power perspective on, 100, 107
egalitarianism, 144
Egypt, 229
Eichacker, Nina, 42
Ejército Zapatista de Liberación Nacional (EZLN), 148–50, 156, 168, 174n, 176, 181, 181n
eldercare, public-power perspective on, 100
Elders Climate Action, 15
electrical grid
architecture of, 234n
capitalists' dependence on, 104
coordination problems of, 55, 56–57
labor power and, 103–04, 105
nationalization of, 97
pumped-storage hydropower and, 88–89, 90, 92, 94
electricity-generation system
boards overseeing public-private systems, 51–52
decarbonization of, 40, 41, 53, 58, 59, 68
electrical grid's coordination problems, 55, 56–57
long-distance high-capacity transmission lines and, 57
marketcraft and, 58, 91–92
potential energy converted to electrical power and, 88
public-power advocates on, 95, 98, 100, 249
regulatory schemes of, 55, 55n
US government intervention and, 91
electricity market, 55–56, 57, 58
electric school buses, 55
electric scooters, 69
electric vehicles (EV)
aftermarket revenue of, 80, 82
battery cell components and, 65
battery industry and, 64, 65
capital raised by manufacturers, 70
charging network for, 53–55, 57, 58, 62
China and, 53, 54, 64–65, 65n, 66
demand for, 69
foreign direct investment in, 64
green transition and, 71, 72, 82
labor power and, 104
leased electric vehicles as tax-credit eligible, 64
marketcraft and, 41, 53–55, 66, 68–76, 108
production of, 94

electric vehicles (EV) *(Cont.)*
public-power perspective in planning of, 96, 96n
sales in designated market areas, 82
tax credit program for, 64
tax rebates for consumers and, 43
transition progress and, 80, 104
US support for manufacturers, 63, 64, 68, 80–81
elemental metals, 42, 42n
El Mozote massacre, 188
El Salvador, 159, 187, 220, 247
embodied cognition, 213n
Emergency Trust Fund for Africa, 229–30
Emergency Workplace Organizing Committee (EWOC), 105
emissions
acceleration of, 77
air travel and, 50
carbon tax proposals, 11
coal emissions, 51–52
fight against, 232
fossil capitalism and, 9
fossil fuel emissions, 6, 7, 74
greenhouse gas emissions, 6, 40–41, 42
mitigation of, 62
in production phase, 9n
relocation of, 73
transferred emissions, 5, 6
Employ America, 42n
Endnotes collective, 131
energy storage, forms of, 87–93, *89,* 94
energy-transition technologies, 41, 43, 47, 70
English Channel, 207
Enlightenment, 117
Environmental Defense Fund (EDF), 5, 6
environmental degradation, 23, *23,* 204
environmental impact reports, 44
environmentalists, 133–34, 197, 198
environmental policy, 11, 11n, 44, 56, 98, 202
environmental protection, 75
Environmental Protection Agency, 44, 51, 55
environmental remediation, 106
environmental standards, 5
equality. *See also* inequality
communism and, 180, 182
gender and, 129
Eritrea, 220
escaped slave communities, 147, 156
Escondida copper mine, Chile, 119
ethics, and arson, 171, 177, 178
Ethiopia, 22, 220
European Central Bank, 176
European Commission, 176
European peasant life, reconstitution of, 141, 155
European Union (EU)
airports ending support for shorter flights, 49
Combined Charging System in, 54
decarbonization policy of, 63
direct investment in electric vehicles, 64
electric vehicles exported to, 65
externalization of border problems, 63, 164
fortress mentality toward climate refugees, 63
migration and, 207, 229–30
neocolonial structural adjustment and, 176–77
ocean extraction and, 216
European uprisings of 1848, 185
everyday life differences, public power and, 107–12
exchange of goods and services, history of, 34
externalized costs, 11, 11n
Extinction Rebellion, 199n
extractive practices
agriculture and, 209–10, 212–13
inequality generated by, 141
ocean extraction, 216–17
oil reserves as unextractable, 4, 5
robber-extractionists, 126
Exxon, 20–21, 82n
ExxonMobil, 6

Facebook, 51
fairness, 238
fair play, rhetoric of, 35
Farahani, Golshifteh, 108
farmers markets, 59
fascism, 25, 179, 237, 254
fears
of disorder, 239, 253, 253n, 254–55
of partisans, 238–40
Federal Energy Regulatory Commission (FERC), 55, 57, 115n
Federal Trade Commission (FTC), 50–51
Federici, Silvia, 190, 190n
Felício, Erahsto, 156–57
Feltrin, Lorenzo, 24n
feminism, 146, 173
Ferdinand, Malcom, 145, 187
FETE (from each, to each), 143–46, 150–53, 180, 249, 252
Field, Alexander J., 71–72, 71n
Fields, Barbara J., 35
Fields, Karen E., 35
fish and seafood production, 211, 216–19, 217n
fish population, 216, 217, 218
Fitch, 59
floods, projections for, 197
Floyd, George, 166–67, 178
FMO (Dutch bank), 120
Follow This, 6
Food and Agriculture Organization of the United Nations, 154, 155–56
food sovereignty, 149, 158
food systems
agrifood systems, 70

INDEX

collapse of, 211
communism and, 143
concentration of agricultural output, 210
extractive agriculture and, 209–10, 212
farming capital and, 210–11
fossil fuels and, 208–09, 210, 213
of Indigenous peoples, 154–58
input sellers and output buyers, 211
land and, 208–13
LED-powered warehouse farms, 209
monocrop exports and, 211–12, 215
natural disasters and, 225
resiliency of, 59
ultraprocessed foods, 215
waste in, 69, 210
weather shocks and, 210
world market and, 211
food-waste reduction, 69
Ford, 54, 65
Forensic Architecture, 207
forest fires, projections for, 197
forest production, subsidies for, 99–100
Fosen Vind, 120–21
fossil capital, 9–10, 9n, 232
fossil fuel asset stranding, 228
fossil fuel emissions, 6, 7, 74
fossil fuel industry
capitalization of, 7
consequences of, 20
electricity market and, 56
fracking boom and, 37
infrastructure targeted by climate movement, 170
labor market and, 132–33
marketcraft and, 48, 51
nationalization of, 97, 237
pipelines of, 56, 130–31, 237
regulation of, 59
World War II war-materials production and, 71
fossil fuels
air travel and, 48–49
American household fossil fuel consumption and, 13, 68
burning increasing amounts of, 82
capital's investment in, 105, 246, 251
decarbonization replacement plan and, 74
energy storage and, 92
fertilizer and, 160, 208–09, 210
food system and, 208–09, 210, 213
freezing consumption of, 62
government control of, 102
as inanimate substrate made of living things, 12
indirect consumption of, 13, 68
labor power and, 134–35
profits from, 3, 4–7, 18, 24, 44, 48
reducing demand for, 40, 113
reducing value of, 93
reorienting world economy away from, 36
replacement of, 72, 107
rules and conditions devaluing of, 36
as "thick" energy source, 10n
transition from, 134, 227–32
US Department of Defense and, 231
whale hunting and, 74–75
fracking, 37
France, 49, 140, 165, 185, 228
Franco, Marielle, 248
Fraser, Nancy, 15n
freedom
communism and, 145–46, 181
fossil fuels and, 209
of movement, 164
public power and, 101
free market, 244
free trade, 158
French Revolution, 140, 228
Fridays for Future, 15
Friedan, Betty, 108
Friedman, Thomas L., 33–34, 37

Gabor, Daniela, 63, 95
Galápagos Islands, 217
García, Brigitte, 248
Gavidia, Venezuela, 162
Gelderloos, Peter, 165
gender
class position and, 127–29
communal values and, 189–90
communism and, 145–46
domestic equality and, 129
nuclear housework and, 129
state terror against women and, 128, 189–90, 190n
gender freedom, 145–46
generative AI software, 227
Germany, 63, 134–35, 147, 186
Gil, Gabriel, 161
Gilroy, Paul, 164–65
global financial crisis of 2008, 65, 176–77
Global Fishing Watch, 218
globalization, 147–48, 158, 159, 190, 190n, 220–21
global South, 62, 63, 99
global warming. *See also* climate change
acceptance of, 197
decarbonization replacement plan and, 74
deprioritization of, 198
global inflation and, 226n
individual responsibility and, 198
land-food system's vulnerability to, 212, 213
oil companies and, 4
ozone-depleting substances and, 67
GM, 54
Go Nuts Biking, 90

Goodell, Jeff, 205
Google, 50
government deregulation, 222–23
Goya, Francisco, 203
Graduate, The (film), 231
Graeber, David, 147, 150, 158, 190n
gray-market employment, 151
Greece, 164–65, 176–78, 186, 207
green aluminum market, 66
green banks, 36, 44, 45–46
green capital, 246
green capitalist policy, 234n, 251
green energy production and distribution
  boom in, 48
  building means of, 36
  competitiveness of, 51
  environmental restrictions and, 51
  Indigenous peoples and, 120
  marketcraft and, 40
  private investment in, 33–34
  subsidies and, 41–44, 47
  US deficit-spending and, 62
greenhouse gas emissions, 6, 40–41, 42
Greenhouse Gas Reduction Fund (GGRF), 44
green industrial policy, 41, 50
green infrastructure, standards implemented for, 53
green investment agenda, 50
Green New Deal, 33–34, 37, 61–62, 103, 108
green trade war, 64
green transition
  class position and, 99
  electric vehicles and, 71, 72, 82
  equalization of global exchange and, 97
  government direction of, 95
  grid-scale energy storage and, 94
  incentivizing of, 95
  Indigenous peoples and, 120
  investment and, 70
  just transition, 97
  labor market for, 59
  labor power and, 134–35
  liberalism's approach to state power over economy and, 33
  markets and, 52–53
  political opinion of, 92
  public-power perspective on planning, 97, 121–30, 133
  pumped-storage hydropower and, 89, 92
  studies on, 69, 75
  Waffle to the Left's enactment of, 32
Grigoropoulos, Alexandros, 176–77
Grigris (film), 19–21, 22, 62, 139, 141
Grosse Pointe Blank (film), 230–31
Grunwald, Michael, 52
Guatemala, 188, 190n, 206, 207, 220
guayule research, 71, 71n
guerrilla parks, 176
Gutiérrez, Lucio, 118
Gwangju massacre, 188

Haber-Bosch process, 208–09
Haley, Nikki, 81
Hampton, Fred, 186
Hansberry, Lorraine, 235–36
Haroun, Mahamat-Saleh, 19–20, 22
Harper, William Jackson, 109
Harrison Power Station, West Virginia, 250–51
Hawaii, natural disaster in, 223
health-care system
  access to, 205
  in China, 244–45
  public-power perspective on, 99, 100, 131
heat pumps, 43
Hidden Fortress, The (film), 242–43
Higa, Teruo, 159
Hiwassee Dam, North Carolina, 91
Hochul, Kathy, 135
Holmberg, Áslat, 120
homeless encampments, 253
Honduras, 120, 190n, 220
housing
  knowledge-intensive housing, 162
  marketcraft and, 38, 253
  public-power perspective on, 100
  renters and, 79–80
  social housing, 234n
  tent encampments and, 167
How the Grinch Stole Christmas! (film), 77–78
Huber, Matthew, 13, 95, 97, 103–04
Human Rights Watch (HRW), 229
humans
  as animals, 213, 214, 216, 222
  human perturbation of wild animals, 214–15
  planetary crisis and, 213–22, 213n
  reaction to animal behavior, 213–14
hurricanes, 223, 224
HVAC systems, 43
hydroelectric dams, 91, 120
HydroWIRES (Water Innovation for a Resilient Electricity System) Initiative, 92
Hyundai, 64

Ibn Saud (king of Saudi Arabia), 78
Ibsen, Henrik, *An Enemy of the People,* 196–97, 198, 199, 199n, 200
Ignatiev, Noel, 127
immigration, 163–64
Imperial College London, 68–69
income per capita
  environmental degradation and, 23, *23*
  inequality and, *22*
India, 66, 120, 147
Indigeneity
  functional meaning of, 116

social metabolism and, 116, 119, 149, 152, 157, 158
Indigenous and Tribal Peoples Convention of 1989, 116
Indigenous Nahua peoples, 187
Indigenous peoples
communal forms of resistance, 119, 147–48, 150, 189
communal systems of, 150–54, 155
food systems of, 154–58
hunting and fishing practices of, 120
land-based conflicts and, 117, 119–20
land management of, 156
laws of, 152–53, 154
Marxism and, 151, 152
population of, 154
postcolonial massacres of, 189
problem of the Indian and, 113–21
rights of, 148–49
settler institutions and, 117, 118, 119–20
sovereignty of, 118
Tennessee Valley Authority's disturbance of burial sites, 116–17
values of, 116, 120, 121, 152
wilderness and, 215
Indigenous Uyghur peoples, 119
Indonesia
anticommunist campaign in, 188–89
ban on unrefined nickel exports, 97
fisherfolk associations in, 147
fishing villages of, 217n
Marind communities in West Papua and, 215–16
industrialization levels, 22, 217–18, 220–21, 250
inefficiency, fears of, 238–39
inequality. *See also* equality
capitalism and, 16, 21, 122, 126, 141, 163
communism and, 142
fossil fuel consumption and, 62
income per capita and, *22*
Inflation Reduction Act of 2022 (IRA)
climate spending and, 38
coal drawdown accelerated by, 52
direct pay provision of, 101
electric vehicles and, 64–65, 68
green energy push in, 40
Greenhouse Gas Reduction Fund (GGRF) of, 44
private investment in green energy and, 33–34
pro-union provisions in, 245
infrastructure
for clean energy, 227
of fossil fuel industry, 170
green infrastructure, 53
investment in, 70
Infrastructure Investment and Jobs Act, 40
insider trading, 35
interaction, models of
communism and marketcraft, 251–53
marketcraft and public power, 243–46, 256
metastrategy of coherence and, 237, 253–58
public power and communism, 246–51
Intergovernmental Panel on Climate Change (IPCC), 228–29
internal dissidents, 185
International Bird Rescue Research Center, 39
International Brotherhood of Electrical Workers (IBEW), 103–04, 135
International Coffee Agreement, 148
International Energy Agency, 74, 92
International Forum on Feminicides, 189–91
international hierarchy, 122, 124–25, 126n, 131
International Labour Organization (ILO), 116, 148, 226–27
International Longshore and Warehouse Union (ILWU), 247–48
International Monetary Fund, 176
International Organization for Migration, 207
International Renewable Energy Agency, 41
international solidarity, 241, 256
investment
in climate adaptation, 62–63
in energy-transition technologies, 41, 47, 70
green transition and, 70
in infrastructure, 70
private capital investment, 134
private green investments, 227
public investments in decarbonization, 45–47, 47n, 48, 53, 58, 63, 94
in transportation, 70
investment funds
climate-related investment in agrifood systems, 70
marketcraft and, 48, 50
ownership of oil companies and, 5–6, 7
socialization of, 6, 47–48
venture capital investment, 69
Iran, 173, 174, 188, 218–19
Iraq, 173, 218, 220
IRS, 38
ISIS, 174
Italy, 165

Jacobs, Justin, 82n
Japan, 64, 150, 159–60
Jarmusch, Jim, 108
Jenne, Addie, 135
jet fuel, 40, 48–49
Jevons, William Stanley, 75
Jevons paradox, 75–76
Joint Office of Energy and Transportation, 54–55

Kaba, Mariame, 166
Kadri, Ali, 230
Kafka, Franz, 111
Kaua'i Island Utility Cooperative (KIUC), 113–14, 115n

INDEX

Kelton, Stephanie, 37–38
Khalaf, Hevrin, 248
Khan, Lina, 50–51, 51n
Kia, 64
Kichwa peoples, 118
Kim Jong-un, 249–50
King, Martin Luther, Jr., 235, 258
Kizza-Besigye, Anselm, 126n
Klee, Paul, *Angelus Novus,* 258–59
Klein, Naomi, 27, 222
knowledge-intensive agriculture, 162
knowledge-intensive energy, 162
knowledge-intensive housing, 162
knowledge-intensive transportation, 162
Koont, Sinan, 257–58
KRRS, 147
Kulmiye, Abdirahman J., 218
Kuna of Ecuador, 147
Kurdish people, 173–74, 181
Kurdistan Workers Party, 173
Kuznets curve, 21–23, *22, 23,* 24n, 126

labor market
  exploitation of cheap labor, 98, 142, 143, 148, 151, 221–22, 227, 228, 230, 231
  fossil fuel industry and, 132–33
  housewifized labor, 128–29
  inclusion through exclusion and, 132n
  international value of, 125
  interventions in, 51
  labor's terms of trade (LToT), 124–25, 126
  marketcraft and, 59
labor power
  assumption of political unity within, 129–30
  bargaining for the common good and, 105
  bargaining within capitalism and, 130, 132
  brushfire organizing and, 105
  capital and, 134, 175, 230
  capitalists and, 9, 9n, 10, 77–78, 103–04, 106, 121, 122, 130, 217, 230
  care and service workers and, 100, 104–06
  class position and, 9, 10, 16, 78, 105, 106, 126
  decarbonization and, 104
  divisions within, 126
  electrical grid and, 103–04, 105
  emergency scenarios and, 262
  environmentalists and, 133–34
  exhaustion of, 15
  exploitation of, 10, 98, 100, 109, 122–23, 130, 132–33, 226–27, 228, 230, 231
  international solidarity and, 247
  legitimacy through elections, 179–80
  personnel problems and, 18
  pink collar labor and, 104–05
  proletarianization and, 126–27
  public-power perspective on, 100, 103–07, 110–11, 122, 133, 134–36, 246–51
  realization of, 8–9, 9n, 17
  selling time below value, 10
  social justice and, 135
  strike in Greece and, 177
  superexploitation of, 129, 131, 151
  unequal global division of labor, 123
  workers renting opportunity to work, 10
Lake Hodges plant, San Diego County Water Authority, 89
La Matanza (the Massacre), 187–88
land
  common land, 141
  food systems and, 208–13
  land-based conflicts of Indigenous peoples, 117, 119–20
  land management of Indigenous people, 156
  land-use restrictions, 56
  urban land occupations, 165
  wildland-urban interface, 215
Latin America, 119, 256
Latour, Bruno, 169
La Via Campesina, 158
lawfare strategy, 51
left, the. *See also* liberals and liberalism
  absolutism as internal threat, 240
  alt-right and, 254–55
  capitalism and, 61, 251–52
  challenges faced by, 260–61
  on complicity, 239
  disaster councils and, 258–62
  international solidarity and, 241
  labor power and, 248, 250
  left-liberal politics on climate crisis, 25
  metastrategy and, 254, 257
  organs of social coherence, 260
  police and, 241
  radical left, 32, 177, 178–79, 247
  support for Sandinista government in Nicaragua, 255–56
  voting support of, 242, 242n
  women's collective self-liberation and, 241
legitimacy
  of communism, 169, 179–84
  electoral legitimacy, 179–80
  new system of, 178
Lego Group, 79
leisure, universal lifestyle of, 33
Lenca peoples, 120
Lenin, Vladimir, 186, 251
Li, Minqi, 124–25, 129
liberals and liberalism. *See also* left, the; progressive liberals
  civil disobedience and, 170
  on collective problems, 26, 27
  fossil fuel transition and, 229
  Green New Deal and, 33–34
  neoliberalism's critique of, 32–33
  political sphere and, 228
  rise of, 72–73

INDEX

liberation praxis, 178
Liebknecht, Karl, 186n
Life, Value's domination over, 212
lithium, 42, 96–97
lithium batteries, 89, 96
living standards, 62
localism, 145
Local Meat Capacity Grant Program, 59
Lockheed Martin, 231
Longfellow, Brenda, 16
Los Bandoleros (film), 183
Lozano Lerma, Betty Ruth, 189
Luisa Cáceres Commune, 256
Lula da Silva, Luiz Inácio, 246–47
Lumumba, Patrice, 188
Luxemburg, Rosa, 185–87, 186n
Lyon, France, 185

Macas, Luis, 118
McKinsey & Company, 22–23, 81–82
Macron, Emmanuel, 63
Maine Public Utilities Commission, 58
Malawi, 245
Malaysia, 66
Mali, 219–20n
Malm, Andreas, 9n, 23–24, 24n, 170
Manchin, Joe, 38, 64
"Manifesto of the Equals" (1796), 140–41, 172
Manji, Rizwan, 109
manufacturing sector
  capital raised by, 70
  creation of jobs, 41
  low-paid labor and, 98
  planned obsolescence and, 59
  postwar automobile manufacturing, 71
  school bus manufacturers, 52
  US support for electric vehicle manufacturers, 63, 64, 68, 80–81
Manzanar concentration camp, 71n
Mao Zedong, 188
Maori of New Zealand, 147
Mariátegui, José Carlos, 117, 119, 141, 150
Marind communities, 215–16
marine ecosystems, 217
Marini, Ruy Mauro, 129
Mariutti, Enrico, 73
marketcraft
  administrative marketcraft, 51
  air travel and, 48–50
  Alice's Farm and, 31–33, 107
  capitalists and, 48, 76–83
  challenges to, 60
  class compromise and, 103
  communism and, 251–53, 256–58
  compensatory action and, 62, 63
  countermarketcraft, 256
  dark marketcraft, 48–49
  decarbonization strategies, 40–43, 66, 238
  derisking strategy and, 63–64
  electric vehicles and, 41, 53–55, 66, 68–76, 108
  emergency scenarios and, 254, 262
  empathy for consumers and, 43
  environmental restrictions and, 51
  green carrots and, 39–47, 52
  greencetera, 52–60
  green sticks and, 47–52
  housing markets and, 38, 253
  incentive structure and, 92
  internationalist strategies, 67
  legitimacy through elections, 179–80
  market outcomes and, 37–38
  nationalist strategies, 61, 63–65, 66, 97
  national marketcraft, 245–46
  outcompeting capitalist system and, 142
  permitting and, 52, 55, 56
  policy briefs and, 107, 110
  politics and, 39–40, 43, 44, 144
  progressive strategy of, 76
  public power compared to, 93, 95, 98, 107–08
  public power interaction and, 243–46, 256
  public-private marketcraft, 131–32
  regulation/deregulation binary and, 55, 58
  rejection of government intervention and, 38
  role of automobiles in, 70–71
  sabotaging of, 231
  shock doctrine and, 223
  socialism compared to, 37
  strategic action and, 234, 237, 238, 240
  strategy of, 32, 34–41, 58, 60
  supplementation of, 83
  taxes and bans crafting capital markets, 50
  technocratic level of disaster response and, 110
  "throw a steak" tactic and, 36–37
  us against the world and, 61–67
market outcomes
  marketcraft and, 37–38
  market-fundamentalist perspective of, 11–12, 26, 34, 37, 51, 53, 197–98
markets
  automation of production system and, 33
  collective interest and, 34–35
  competitiveness of, 50–51
  denaturalization of, 39–40
  deregulated markets, 37
  effect of long-term investment in clean energy on, 36, 42
  efficiency of free market, 201
  electric vehicle charging network and, 53–54
  as function of public policies, 34
  government standards and, 33
  green transition and, 52–53
  as institutions, 35
  maintaining rules of, 175
  metabolic management of, 35
  policymakers' attempts to improve, 11

markets *(Cont.)*
  public investment authority and, 47
  rule-making polity and, 34, 53
  spot and futures markets for minerals, 42n
  subordination to society, 35
  taxes and bans crafting, 50
  values and, 34–35
Martí, José, 258
Marx, Karl, 158n
Marxism, 151, 152, 173
mass mortality events (MMEs), 214–15
Matanza, La (the Massacre), 187–88
Mau, Søren, 16–17
Mauritania, 219–20n
meat and poultry industry, 220–22, 230
medical inhaler manufacturers, 67
medical system, public-power perspective on, 98, 99, 100
Merchant, Carolyn, 141–42, 145
Merryman, Kenneth, 235
Mészáros, István, 14, 143–44, 240
metastrategic pragmatism, 27, 101, 195, 201, 254
methane, 213
Mexico
  capitalism in, 148
  US border policy and, 63, 164, 206–07
  violence against women in, 190n
  workers recruited for American meat and poultry industry, 220
  Zapatistas of Chiapas, 148–50, 156, 168, 174n, 176, 181, 181n
Meyerson, Harold, 50
microhydropower, 88
Microsoft, 51
Midwestern Corn Belt, 210, 211
Mies, Maria, 127–28
migrants and migration
  agricultural products and, 215
  asylum protections and, 206, 207–08
  border imperialism and, 163–65, 229
  climate change and, 63, 164, 200, 204–08
  criminalization and racialization of, 163, 165, 166, 167–68
  globally planned migration, 228
  motivations and, 205–06
  nationalism and, 229–30
  planetary crisis and, 200, 204–08
  solidarity movement and, 176
  US internal climate migration, 207n
Migration Policy Institute, 164
military-industrial complex, 231–32
Millner, Naomi, 159–60
minerals, supply of, 42–43, 42n
mine-waste refining, 43–44
mining industry
  artisanal mining and, 151, 212, 219–20n, 247
  battery-materials miners, 81–82, 93
  planetary crisis and, 209
  public power and, 96–97, 119
mining workers, 104, 151
Minneapolis, Minnesota, 166–67
Minnesota, 162
Mitchell, Timothy, 78, 180, 199
Mkandawire, Thandika, 245
mobile capital, 63–64
monopolization, capital and, 50
monopoly power, growth of, 80
Monsters, Inc. (film), 47
Montreal Protocol, 67, 79
Moody, Kim, 106
Moody's, 59
moral humanism, 171
moral imperatives, 240
Mosaddegh, Mohammad, 188
mountains
  batteries built into, 87–88
  pumped-storage hydropower and, 88, *89*
Movimento dos Trabalhadores Rurais Sem Terra (MST), 156–57
Moyo, Sam, 150–51, 156
mutual aid, 252
Myanmar, 220

Nā Kiaʻi Kai (sea protectors), 114
Napoleon, Val, 152–53, 172
narcotics industry, 19
National Automobile Dealers Association (NADA), 80–82
National Beef factory, 220
National Capital Management Corporation (NCMC), 46–47
National Charging Experience (ChargeX) Consortium, 54
National Environmental Policy Act (NEPA), 44, 56
National Homelessness Law Center, 167
National Intelligence Council (NIC), *A More Contested World,* 204, 206
National Investment Authority (NIA), 45–47, 46n
National Investment Bank (NIB), 45–46
nationalism
  border policies and, 164, 229
  marketcraft and, 61, 63–65, 66, 97
  transition and, 227–32
  win-lose contests and, 25
national socialism, 127
National Union of Mineworkers, Marikana, South Africa, 248
Native Hawaiian cultural rights, 114–15
Native Potato Project, 162
natural disasters, capital accumulation and, 222–27, 231
nature
  capitalism threatening relationship with, 15

communism and, 143
ecosocial metabolism and, 150
exploitation of cheap nature, 74, 77, 98, 131, 142, 143, 148, 151, 210–11, 212, 214–15, 217, 224, 226
Navajo Nation, 115n
needs. *See also* public wants and needs
communism's definition of, 143–44
shared needs, 48
societal needs, 95–96
Neel, Phil A., 259n
Nelly, 80
neoliberalism, 32–33
neo-Luddism, 253
Nepal, 120
Netherlands
airports ending support for private flights, 49
derisking strategy and, 63
electric vehicles and, 53
renewables subsidized with household natural gas taxes, 123n
New Deal, 91, 108, 110
New Sanctuary Movement, 165
New York Power Authority's (NYPA), 102
Nicaragua, 220, 255–56
nickel, 42, 73, 97
Niemöller, Martin, 191
Nigeria, 117
Nipper, Mads, 79
Nishnaabeg peoples, 151–52
No Border network, 165
Nolan, Hamilton, 131
No More Deaths, 165
noncompete agreements for workers, 51
non-renewable energy, price of, 41
North American Charging Standard (NACS), 54
North American Free Trade Agreement (NAFTA), 148, 220
northern spotted owl, 133, 133n
North Korea, 249–50
North Saskatchewan River, oil spill of 2016, 12
Norway
electric vehicles and, 53
Sámi peoples and, 117, 120–21
NOTOFLOF (no one turned away for lack of funds), 252
Nous Sommes la Solution (NSS), 158–59, 168
noxious deindustrialization, 24n
nuclear energy production, women handling health costs of, 129
nuclear power plants, 56–57

Oak Ridge National Laboratory, 89
Obama, Barack, 37, 51, 118, 198
O'Brien, M. E., 145–46, 184
Öcalan, Abdullah, 168, 173–74
Ocampo, Omar, 49
Ocasio-Cortez, Alexandria, 37
Occupy movement, 176
ocean extraction, 216–17
Oceti Sakowin peoples, 118, 170
offshore humanism, 164–66
Ogawa Productions collective, 150
Ogoni peoples of Nigeria, 117
oil and gas sector
capitalization of, 7
contraction of, 3
investments in exploration, 40
jobs provided by, 21
mergers and acquisitions in, 5
reserves of, 28
shale oil boom, 20
spot and futures markets for, 42n
value of, 59
oil industry
capitalism and, 4, 5, 6–7
capitalists and, 78
in Chad, 19–20
climate restrictions on, 3–5
democracy as oil and, 78
denial-and-delay complex and, 198
extraction and, 228
government control of, 102
petroleum as feedstock and, 72
in postwar economy, 71
shareholders of, 5–6
US and world economies intertwined with, 7
use of subsidized ethanol, 211
oil palm industry, 215–16
oil reserves, as unextractable, 4, 5
Oil-Value
labor power and, 8–9
marketcraft eliminating link between, 35
meaning of, 8–15, 8n, 16
realization of, 8–9
Oil-Value-Fish chain, 219
Oil-Value-Life chain
breaking link of, 35, 40, 234, 260
communism and, 142–43, 163
food system and, 210
intervention in, 12
persistence of, 13
planetary crisis and, 195, 196, 201
public power and, 93, 100
Oil-Value-Sheep chain, 219
Omarova, Saule, 45–47, 46n
Operation Condor, 188
Operation Warp Speed, 98
Orbán, Viktor, 172
organopónicos (urban farming technique), 257–58
Ørsted (offshore developer), 79
ozone-depleting substances (ODS), 67
ozone layer, 67

Pachakutik Movement, 118
Pacific Northwest, 133

Pakistan, 224
Palacios, Yarsinio, 205–06
Palladino, Lenore, 107
Papa para la vida, no para el capital (Potatoes for life, not capital), 161
Parenti, Christian, 229
Paris Agreement, 4, 81
Paris Commune, 185
Partido de Trabalhadores, 246
partisans. *See also* communism; marketcraft; public power
  building relationships between, 260, 261–62
  coherence of, 234–38, 234n, 240–43, 247, 253–62
  emergency scenarios and, 254–55, 261–62
  fears of, 238–40
  models of interaction, 243–57
  mutual understanding among, 239
  role of, 26–27, 234, 236, 241
  shared values of, 254
Patel, Priti, 207
Paterson (film), 108–10, 108n
Pāti Māori, 118
patriarchy, 122, 127–29, 189
Paul, Mark, 42
Penrod, Emma, 57–58
permits, for construction projects, 52
Persian Gulf, 219
personal responsibility, 100
petroleum
  derivatives of, 39
  as feedstock, 72
  history of use of, 12
Philippines, 190n
Philpott, Tom, 210
photosynthesis, process of, 236–37
photovoltaic energy, carbon intensity of, 73–74
Pimenta, Joana, 182–83
Pirani, Simon, 13
planetary crisis
  animals and workers, 213–22, 213n
  anthropocentric view of, 216
  climate and migration, 200, 204–08
  conscious consumption and, 197
  coordinated response to, 201
  dancing pairs and, 203–04, 205, 206, 213
  democracy and, 199–200
  denial of scope of, 195–97, 200, 202
  diversity of thought on, 236
  economic sanctions and, 200
  environmental policy and, 202
  immediacy of, 197
  inadequacy of solutions apparatus, 200
  individual choices and, 197–98, 200, 201
  insurrectionary globalist conspiracy against, 201, 202
  land and food, 208–13
  local crystallizations of, 261
  market fundamentalist strategy and, 197–98
  natural disasters and profits, 222–27, 231
  Oil-Value-Life chain and, 195, 196, 201
  overwhelming nature of, 233
  paths for progress, 236–37
  policy solutions for, 204, 234–35, 234n
  quantum walk strategy, 236–38, 260
  realism and, 233–34
  several paths taken at same time, 236–37
  shifting nature of, 195–202
  social metabolism and, 201, 239–40
  solvable problems and, 195–97
  strategic action on, 234–35, 236, 237–38
  transition and nationalism, 227–32
  Value and, 235, 261
  world-scale nature of, 200
planned obsolescence, 59, 234n
plastics, 68
Plato, 203
Platonov, Andrey, 175n
Pō'ai Wai Ola, 115
police, 166–68, 178, 179, 223, 240–41, 249, 254
Political Economy Research Institute, 59
pollution
  automobiles and, 68–69
  as capitalist externality, 11, 11n
"pool party progressivism," 90n
Popular Front for the Liberation of Palestine, 170
postcolonial exploitation, 62
postmodern nihilism, 26
Postone, Moishe, 131
poverty, 21, 62
power, building of, 241–42
power storage, domestic utility-scale power storage, 87
predatory competition, 50
premature industrialization, 22
preventive medical care, 99
primary mediations
  as absolute needs, 14, 143, 183
  conditional access to, 17–18, 183, 239, 252
  sharing of, 237, 254
Princeton University, 69
principles, fidelity to, 242–43
private equity, 51, 70
private flights, 48–50
private/public division, 145
privatization, 222–23
Procter & Gamble, Dawn soap for cleaning animals after oil spills, 39
production
  automation of, 33
  capacity of, 51
  capitalists owning means of, 9, 16, 76–77, 98, 102
  class power balance and, 241–42
  communism and, 145, 152

conditional access to means of production, 17–18
consumption separated from, 13, 14
democratic goals of, 40
efficiency of, 75
emissions in production phase, 9n
food systems and, 210, 211, 212
labor power and, 106, 130
nationalization compared with public power, 97
for private profit, 102
production for production's sake, 34, 40, 133, 143
public-power perspective on, 101, 102, 140
regenerative production, 99
scale of, 218
social appropriation of means of production, 93–94
social problems resulting from, 141
superexploitation of earth's systems and, 142
un/equal exchange between classes as basis of production of value, 9, 9n
unproductive/productive binary and, 152
of Value, 123
value production for value production's sake, 11, 14, 15, 17, 24
waste products of, 141
zero-waste production, 234n
profit rate
of capital, 237
decline in, 228
profits
from fossil fuels, 3, 4–7, 18, 24, 44, 48
natural disasters and, 222–27, 231
progressive liberals
capitalism and, 14, 24, 232
disaster responses guided by values of, 261
on market regulation and incentives, 33
planetary crisis and, 195, 234
progress narratives
Alice's Farm and, 31–33, 107
Walter Benjamin on, 258–59
collective problems and, 26, 27
of liberals, 32
revolutionary change and, 259
proletarianization, 126–27, 184
property claims, of capitalism, 169
property ownership, power of government reserved for, 180
property relations, 17
Prospectors & Developers Association of Canada (PDAC), 81–82
protest camps, 184
public banking, 234n
public credit-rating agency, for asset evaluation, 59
public equity funds, 45
public life, public power and, 98–102
public policies
market actors and, 53
marketcraft and, 58
markets as functions of, 34
public power
accumulation of, 102
affirmation trap and, 130–36
antilocalist strain of, 115
building of, 135
bus as signature vehicle of, 108
capital as opposite of, 102–07
capitalists and, 94–95, 96, 97, 99, 100, 102, 246
care and service workers and, 100, 104–06
challenges to, 114
class conflict and, 103
class position and, 102, 103, 105–06
colonialism and, 114–15, 116, 121
communism and, 140, 246–51
cycle of accumulation, 103
dedication to security, 239
democracy and, 94–95, 102, 103, 121–22, 144
differences in everyday life and, 107–12
electricity-generation system and, 95, 98, 100, 249
emergency scenarios and, 254
everyday life differences and, 107–12
financing and planning capacity of, 101
freedom and, 101
in Greece, 177
green transition and, 97, 121–30, 133
internationalist advocates, 125
labor power and, 100, 103–07, 110–11, 122, 133, 134–36, 246–51
legitimacy through elections, 179–80
marketcraft compared to, 93, 95, 98, 107–08
marketcraft interaction and, 243–46, 256
outcompeting capitalist system and, 142
Paterson and, 108–10
people's control over modern technology and, 93–97, 99
planned fairness and, 111
problem of the Indian and, 113–21
public-interest mandates of, 101
public life and, 98–102
Raccoon Mountain Reservoir and, 90–93, 94
shock doctrine and, 223
socialistic agenda of, 94
social metabolism and, 96, 97, 100, 101, 111, 114, 115, 122, 123, 128–29, 136, 152
social planning and, 96, 96n, 97, 107–12, 121, 123, 128
Storm and, 110–12
strategic action and, 234, 239, 240
superexploitation and, 121–30, 131
transition to, 95, 114
values and, 115, 123
working-class people and, 103, 104–07, 122, 123, 128

Public Power NY campaign, 101–02
public universities, government-funded labs at, 52
public wants and needs
  government financing with subsidies, 40, 41–44, 48, 52, 95, 103
  marketcraft and, 51
  private capital fulfilling, 38
Pueblo a Pueblo, 160–61, 182
pumped hydroelectric storage facility, 88, *89*
pumped-storage hydropower (PSH), 88–93, *88*, 94, 113, 114–15
Putin, Vladimir, 172
PwC (formerly PricewaterhouseCoopers), 69

Queirós, Adirley, 182–83

Raccoon Mountain Reservoir, Chattanooga, Tennessee, 90–93, 94, 97
racecraft, 35
racialism, 25, 126–27
racism, 122, 235
radical left, 32, 177, 178–79, 247
radioactive waste, 129
rare-earth batteries, 89, 93, 96
Ray, Satyajit, 198–99
RCEE (ruthless criticism of everything that exists), 153, 154, 157, 159, 159n, 180
realism, 27, 233–34
Reconstruction Finance Corporation, 45n
Re.Co.Sol, 165
Red de Municipios de Acogida de Refugiados, 165
"reduce, reuse, recycle," 198
refining workers, 104
regenerative production, public-power perspective on, 99
regulation/deregulation, conceptual binaries of, 35, 55, 58
religious values, 198–99
Renault, Marion, 215
renewable energy sources
  development of, 75, 107, 113–14
  discharging of, 87
  storage of, 93
reproduction
  food systems and, 210
  private realm of, 145–46
Republic Windows and Doors factory, Chicago, 250
Rice Lake, Minnesota, 162, 162n
rideshare platforms, 76
Right to Repair Act, 58–59
Riofrancos, Thea, 118
robber-extractionists, 126
Robinson, Cedric, 126–28
rocket weapons industry, 231
Rodrik, Dani, 22
Roosevelt, Franklin, 43, 91, 245
Roosevelt Institute, 40–41, 42
rose theory, 168–74, 179, 180, 205
Rousseau, Henri, *Les joueurs de football*, 259–60
Royal Dutch Shell, 3–4
rubber
  guayule research and, 71, 71n
  synthetic rubber program, 71–72
Russia, 38
Rwanda, 164

S&P, 59
Sabel, Charles, 67
sacrifice zones, 24n, 241
SAE International, 54
Salman (king of Saudi Arabia), 78
Sámi peoples of Norway, 117, 120–21
Sammon, Alexander, 80–81
Sam Moyo African Institute of Agrarian Studies, 150
San Andrés Accords, 148–49
Sanders, Bernie, 37–38
Sanders, Deion, 80
San Diego County Water Authority, 89
Sandinista government, Nicaragua, 255–56
Sankara, Thomas, 188
Saudi Arabia, 78
Saval, Nikil, 43
school bus manufacturers, 52
Schwarzbard, Sholem, 172
Schweizer, Roman, 231
Scientific Assessment Panel, 67
scientism, 15–16
sea-level rise, projections for, 197
Seattle, Washington
  Capitol Hill Autonomous Zone (CHAZ) occupation and, 178–79
  World Trade Organization meeting in, 247–48
second-order mediations, 14, 17
Segato, Rita Laura, 188
self-defense, 168–69, 172
self-interests, convincing people to act against, 16
semiconductor fabrication, US production of, 40
semiproletarianization, 150–51, 158, 163, 212, 216
Senegal, 158–59, 229–30
service sector, 98
settler institutions, 117, 118, 119–20
Seuss, Dr., 77
Seventh Generation, 39
sexual freedom, 145–46
shared needs, 48
shareholders, 5–6, 11
sharing economy, 75–76
Sharpeville massacre, 188
sheep, 219, 220
Shell oil company
  green energy boom and, 48
  Ogoni peoples and, 117

profits from fossil fuels and, 3, 4–7, 18, 24, 44, 48
sky scenario and, 4
Shiva, Vandana, 224
Shoah, 186
Sierra Club, 51
Sierra Nevada, 88
Simpson, Leanne Betasamosake, 151–52, 187
Singer, André, 246
sky scenario, 4
slave industry
capitalist history of, 72–73
coerced fishing labor and, 217
ending of, 187–88
modern slavery, 227
Social Democratic Party, 186
social housing, 234n
socialism
on collective problems, 26, 27
history of, 185–86
Indigenous peoples and, 151
marketcraft compared to, 37
social metabolism
breakdown in, 259n, 261
capitalism as form of, 14–15, 17, 18–19, 24, 94, 125, 131, 141, 143, 175, 180, 190, 215, 239, 261
communism and, 140, 141, 142, 144, 145, 152, 155, 160–62, 163, 175, 181, 249, 258
decarbonization and, 74, 201
defense of, 140
democratic confederalism and, 173
economic power and, 17
ecosocial-metabolic perspective, 150, 160–62, 163, 200, 208–09, 228, 249
Indigeneity and, 116, 119, 149, 152, 157, 158
planetary crisis and, 201, 239–40
property relations and, 17
public power and, 96, 97, 100, 101, 111, 114, 115, 122, 123, 128–29, 136, 152
self-defense and, 168
social planning
capitalists and, 97, 102
communism and, 152, 162, 171, 177–78
public power and, 96, 96n, 97, 107–12, 121, 123, 128
social spending, cuts to, 222–23
societal needs, 95–96
Société des Hydrocarbures du Tchad, 24
society, planning of, 59–60
solar assembly industry, US expansion of, 66
solar energy
environmental variables and, 87
extreme weather and, 227
flows of, 10n
food systems and, 208, 209
Indigenous peoples and, 159
labor power and, 107
land-use restrictions and, 56
permitting and, 56
saving of, 88
Solyndra and, 37, 41, 42, 64, 66
transmission lines and, 53
solar installation industry, 66, 104
solar-panel market, 66, 69, 73–74
solar-plus-storage battery setups, 113
solar-powered water cycle, 88
sole proprietorships, in capitalist system, 9n
Solyndra, 37, 41, 42, 64, 66
Somalia, 218–19, 220
Song, Lisa, 99
Sonko, Mariama, 159
Sonoran Desert, 165
South Africa, 134–35, 247
South America
anticommunist repression in, 188
escaped slaves communities in, 147
negative agricultural efficiency and, 225
South Asia, 124, 125
Southern Off-Road Bicycle Association, 90, 90n
South Korea, 64
sovereign debt crisis, 176–77
Soviet Union, 186, 218, 257
Spade, Dean, 252
Spain
airports ending support for shorter flights, 49
anarchist groups in, 147, 165
squatting, 165, 176, 253
Sri Lanka, fisherfolk associations in, 147
standard of living, basic standard of living for everyone, 93
Starbucks, 105, 131
Statkraft, 120
status quo, reinforcement of, 99
Stewart, George R., *Storm,* 110–12
Stockton, Nicholas, 219
Stop Cop City campaign, Atlanta, 255
Stroehlein, Andrew, 229
Studio Ghibli, 31
sub-Saharan Africa, 125
Sudan, 219, 220
Sum Pak project, 158–59, 159n
sumptuary taxes, 49
superexploitation
of earth's systems, 142
labor power and, 129, 131, 151
public power and, 121–30, 131
supply-side industrial policy, 36, 43, 48
survival, means of, 146
SUV dealership arson, 171
Syria, 165, 173, 174
systematic runting, in pork industry, 221–22

Tabuchi, Hiroko, 5
Tanzania, 22
Technical Options Committees (TOCs), 67

INDEX

technocrats, 144
technological solutionism, 99
technology
  agroecology and, 159
  public power and, 93–97, 99
  types of, 96
Technology and Economic Assessment Panel, 67
techno-utopianism, 26
Teia dos Povos (TdP), 156–57, 182
temperature changes, varied effects of, 74
Tennessee Valley Authority (TVA), 91, 93, 94–95, 97, 101, 107, 116–17, 245
Terán, Manuel Esteban Paez "Tortuguita," 255
Terzi, Alessio, 33
Tesla, 54, 65, 113
Thailand, 66, 217
thermal divide, 205
Third World, 63, 176–77
Thomhave, Kalena, 49
Thunberg, Greta, 170
tire pollution, 68
Todd, Zoe, 12
Torres, Bernabé, 161
transferred emissions, 5, 6
transnational capital, 63, 64
transportation
  decarbonization of, 41, 53, 58
  investment in, 70
  public-power perspective in planning of, 96, 97
tree planting, 99–100
Trouble in Paradise (film), 223–24
trucking industry, 52
Trump, Donald, 66, 172, 206, 254
TSMC (Taiwanese firm), 125
Turia, Tariana, 118
Turkey
  China and, 66
  EU border policy and, 63, 164, 207
  Marxism and, 173
  ocean extraction and, 216
Tutuola, Amos, 18n

Ulyanov, Aleksandr Ilyich, 172
Unification Church, 172
United Auto Workers, 104, 131
United Electrical, Radio and Machine Workers of America (UE), 105
United Kingdom
  anarchist groups in, 147, 165
  Chartist movement in, 185
  externalization of border problems, 164
  migration and, 207
  social divisions within, 126–27
UNICEF, 226
United States
  anarchists in, 165
  anticommunist repression and, 188, 188n
  border imperialism of, 164, 165, 166
  climate strategy of, 68
  communist movement in, 186
  externalization of border problems, 63, 164, 206–07, 208
  fortress mentality toward climate refugees, 63
  internal climate migration in, 207n
  international hierarchy of capitalism and, 124, 125, 126n
  lines of exclusion within, 166–67
  ocean extraction and, 217
  private air travel and, 49–50
USA Cycling, 90
US Department of Agriculture, 59, 71n, 225
US Department of Defense, 231–32
US Department of Energy, 54, 87, 92
US Department of Transportation, 54
US government
  direct intervention for electrical power, 91
  effect of long-term investment in clean energy on markets, 36, 42
  functions of, 76–77
  relations with domestic capital, 61
United Steelworkers, 245
universalism, 129–30
universal politics, 26
urban land occupations, 165
urban/rural communities gap, 160–61
Uyghur peoples, 119

vaccines, 37
Vallet, Élisabeth, 164
Value
  agriculture and, 209, 210, 212, 215–16
  calculation of, 169
  capitalism and, 116, 123, 133, 139–40, 143, 144, 169, 177, 187, 216, 253
  de-development and, 218, 219–20
  ecosocial-metabolic planning practices replacing, 161
  exploitation and, 220
  FETE in defiance of, 252
  left's alternative solution, 261
  Life dominated by, 212
  meat and poultry industry and, 222
  natural disasters and, 225
  planetary crisis and, 235, 261
  shock doctrine and, 223
  as social mediation, 142
  subordinated to values, 12, 143
  un/equal exchange between classes as basis of production of, 9
  value production for value production's sake, 11, 14, 15, 17, 24
  wildland-urban interface and, 215
value creation, 79
Value-Life link
  capitalism and, 15, 16
  marketcraft and, 44

theft and, 139
Value's domination over Life, 212, 217
Value-Oil-Life chain, breaking of, 35, 140
values
alternative systems of, 15
changes in, 7, 132, 260
communal values and, 139–40, 142–43, 187, 189–90
communist values, 159, 175, 177
cultural values, 198–99
of democracy, 229
disaster responses guided by progressive values, 261
of Indigenous peoples, 116, 120, 121, 152
markets attached to, 34–35
non-Value values, 245
public power and, 115, 123
reevaluation of, 199, 235
shared values of partisans, 254
societal needs and, 95–96
Value subordinated to, 12, 143
Vanguard, 6
Venezuela, 160–62, 182, 200, 220, 256
venture capital, 69
Via Campesina, La, 158
Victor, David, 67
Vietnam, 66
violence, state's monopoly on, 172
Vogel, Steven, 35–37
von Kleist, Heinrich, *Michael Kohlhaas,* 172–73
voting rights, 242

Waffle to the Left, 31–32
wage theft, 226
Walia, Harsha, 163
War on Drugs, 19
War on Terror, 220
Warren, Kawai, 114
waste, in food systems, 69, 210
water cycle, conversion of kinetic energy of, 88
Water Power Technologies Office, US Department of Energy, 92
water sector, investment in, 70
wealth, global distribution of, 61–62
weatherization, grants for, 43
Weather Report (film), 16
weather shocks, 225–26, 227
Weber, Isabella, 38
WeChat app, 54
Weelaunee Forest, Georgia, 255
Wei, Zhang, 244–45
West Africa, 158
West Bank, Israeli settlers and, 155
West Indies, colonial slavers of, 72–73
West Kaua'i Energy Project, 113–15, 115n, 120
West Kaua'i Watershed Alliance, 115
Wet'suwet'en peoples, 170
whale hunting, fossil fuels used in, 74–75
whales, 74, 214
Whitest Kids U'Know, 249
Whole-Home Repairs bill, 43, 79
Wilde, Oscar, *The Picture of Dorian Gray,* 10
wildfires, 224
wildland-urban interface (WUI), 215
Williams, Eric, 73
Winant, Gabriel, 105
wind energy
environmental variables and, 87
Indigenous peoples and, 120–21
offshore wind energy, 79
transmission lines and, 53
wind-turbine market, 66
witchcraft, 35
witch hunts, 189–90
Wolf, Martin, 39
women. *See also* gender
collective self-liberation of, 241
housework and, 127, 128
social costs externalized by, 128–29
state terror against, 128, 189–90, 190n
Wong, Moisés Sio, 257
Woolsey Fire, California, 224
worker co-ops, 252
Workers' Youth League camp, Utøya, Norway, 248
working-class people
balance of class power and, 241–42
capitalism and, 16, 106, 121, 122, 126–27, 148
class struggle and, 230
differences among, 129–30
downward pressure on, 227
exploitation of, 123, 131, 132, 132n
gender divides among, 127
global working class, 121, 122
labor movement and, 106, 246–47
as marketcraft constituency, 234
needs of, 229
private air travel and, 49–50
public power and, 103, 104–07, 122, 123, 128
repression of consumption, 124
self-determination of, 247
social divisions within, 126
well-being of, 239, 240
World Bank, 21, 124, 226
World People's Conference on Climate Change and the Rights of Mother Earth, 15
World Trade Organization, 64, 65, 247–48
World War I, state system devised following, 78
World War II
electrical power for industry and, 91
rubber and, 71–72, 71n
war-materials production analogies, 70–71
World Wildlife Fund, 120

Xiaohuan, Lan, 244

Yanomami peoples, 247
Yazzie, Steven, 153
Yemen
  fishing operations in, 218
  food production in, 211, 212
Yeros, Paris, 150–51, 156
YIMBYs ("Yes in my backyard"), 38
York, Richard, 74–75
Young Farmers Action Group, 150

Zapatistas, 148–50, 156, 168, 174n, 176, 181, 181n
zero-waste production, 234n
Zetkin Collective, 232

CREDITS

Chris Parris-Lamb, The Gernert Company, **Agent**
Elisa M. Rivlin, **Legal Counsel**
Jean Garnett and Alex Littlefield, **Editors**
Morgan Wu, **Assistant Editor**
Gregg Kulick, **Art Director**
Kirin Diemont, **Jacket Designer**
Marie Mundaca, **Text Designer**
Solomon J. Brager, **Illustrator**
Daniel Harris, **Fact Checker**
Ben Allen, **Managing Editor**
Pat Jalbert-Levine, **Production Editor**
Barbara Clark, **Copyeditor**
Katherine Isaacs and Deborah Jacobs, **Proofreaders**
Kay Banning, **Indexer**
Emily Baker, **Production Coordinator**
Stacy Schuck, **Manufacturing Coordinator**
Alyssa Persons, **Assistant Director of Publicity**
Darcy Glastonbury, **Publicity Assistant**
Kayleigh George, **Senior Director of Marketing**

Emma Littel-Jensen, **Marketing Associate**
Lauren Hesse, **Social Media Director**
Jen Patten, **Audio Production Supervisor**
Laura Essex, **Audio Producer**
Sally Kim, **Publisher**
Michael Barrs, **Deputy Publisher**

ABOUT THE AUTHOR

Malcolm Harris is the author of the national bestseller *Palo Alto: A History of California, Capitalism, and the World,* a finalist for the Los Angeles Times Book Prize; *Kids These Days: The Making of Millennials;* and *Shit Is Fucked Up and Bullshit: History Since the End of History.* He was born in Santa Cruz, California, graduated from the University of Maryland, and now lives in Philadelphia, Pennsylvania.